PMBOK® Guide and PMP® Exam Master Guide

Study Guide on the Project Management Body of Knowledge with Practice Test Questions for the Project Management Professional Exam

by Robert P. Nathan

Table of Contents

Introduction

Section I.1: Goals

My primary goal in writing this book is to prepare you for the project management professional (PMP) certification exam. Project managers are the jack-of-all-trades in the growing possibilities of project-related certifications. If you are here instead for a certification in risk management or business analysis or you are a project expeditor wanting to learn more about the craft, we will cover much of your relevant material. This book will go through PMP standards and terminology in-full as regulated by the Project Management Institute. It is up-to-date for 2018-2019. Welcome to the journey!

Project management is a growing, exciting, and diverse field. Project managers work anywhere from small firms leading the same group of a few people over similar projects lasting years, to the largest corporations in the world with massive teams on varied projects lasting weeks. My editors and I have worked to have this book apply to every project manager. The profession of project management emerged in the 1960s based on the realization that whatever the project was on, there were enough similarities to warrant a focus on being a PMP. Whether a project manager spends their days working in accounting or engineering, project success depends more on team-building and leadership than mastery of specific content. Whatever your work background is, this book focuses on project management similarities and your successful PMP certification.

Certification will allow you to increase your marketability, network with other project managers, and improve your abilities to lead a project. To take the PMP exam and get certified, there two requirements. You will need to have logged 4,500 hours in managing a project if you have a bachelor's degree or 7,500 hours in managing a project if you do not have a bachelor's degree. Secondly, you will need to have 35 hours in an approved project management education program. There are many options in the education field. Many aspiring to certification choose an online course to best fit their schedules. This book is meant to connect the dots from your experience on the job and in the classroom to best prepare you for the exam.

As any project manager will tell you, time is the most valuable commodity in any project. This book is designed to make the most of the time you spend preparing for the exam. Rather than being hundreds of pages too long, the following chapters have gone through multiple stages of revision to include only those things that you need to know. The chapters have been designed to be readable, as well as to be scannable if you are flipping through. I have bolded all the terms you need to know. Mastering the jargon is a big part of joining the profession. However, buzzwords like paradigm shift, de-risk, and bandwidth, which do not have a specific meaning in project management, are absent in the following pages. Obvious words like higher management will also be skipped over in the interest of your time.

The structure of this book is based on project management's five **major process groups**: initiation, planning, executing, monitoring and controlling, and closing. These are the five phases every project goes through and should therefore always be on your mind. Focusing on the process groups allows for the book to have a flow: a beginning, middle, and end just like a project. There are five chapters on each of these areas followed by ten practice exam questions. The practice questions hone in on the chapter's most difficult material and make connections between multiple topics.

While there is overlap between the process groups, they are distinct and have their own logic. Project management's ten **knowledge areas**, on the other hand, shape project activities from the first to last day. We have placed the knowledge areas with the appropriate chapters. For reference, the ten knowledge areas are: communication management, cost management, human resources management, integration management, procurement management, quality management, risk management, scope management, stakeholder management, and time management. Becoming fluent in these ten areas is the quickest route to certification.

After the five chapters, there is a glossary and a two-hundred question practice exam. You should go to the glossary if you do not recall a term that appeared earlier in the text. The practice exam mimics the real thing in scope of coverage and question type.

There is no filler in the practice exam or in this book. In addition to mastering the material, success on the exam also depends on test-taking best practices like time management, maintaining your concentration, and process of elimination within answer choices.

Section I.2: Test information

For the PMP exam you are allotted four hours to answer 200 questions, with 25 questions being experimental, unscored questions scattered at random through the exam. The exam is broken down by the major process groups, with the number of questions for each domain comprising the following percentages: initiation: 13%, planning: 24%, executing: 31%, monitoring and controlling: 25%, and closing: 7%.

Section I.3: Study app and more

I have partnered with Quizlet to create an app that features all the terms in the glossary. You can get the Quizlet app in the App Store or Google Play. Search for "Robert Nathan Project Management."

The url for the flashcards is: https://goo.gl/7ZRr41.

This book is my project. I do not consider it closed until you are satisfied. Unlike other books written by teams of corporate writers, I am looking for ways to continually improve my work and approach to project management. Let me know if there is something I can help you with. Email your questions and comments to: robert@pmppeternathan.com.

Additionally, if you found this book helpful, please leave a positive review on Amazon. In an industry dominated by large publishers, make your review count.

Best regards,
Robert Nathan

Chapter 1: Initiating

Section 1.1: To begin or not to begin

Whether a firm decides to take on a project or not, fundamental questions must be considered. Is the potential client a good fit? Is this a project within the firm's area of expertise? How will this project be similar or different from past ventures? What is the timeline and how will that interfere with the firm's other obligations? What are the resources this project needs? Who will lead and staff the project? These are general questions that map onto specific definitions within the project-managing profession. In order to answer these questions, let's take a look at some of the basic terminology that is core to project management:

> - A **project** is an undertaking that has a distinct start and end date. A project's goal is to produce a new good or service. Completing a project will require a combination of inputs: natural resources, labor, capital, and/or entrepreneurship. A project's schedule, product, and resources are its main elements.
> - A **project manager** is responsible for the day-to-day planning, execution, monitoring, and closing of a project. A project manager reports on progress and timetables to project stakeholders, contractors, and higher management within the performing organization.
> - A **project sponsor** is above the project manager within the same organization, sets deadlines, oversees funding, and provides guidance during the project's duration. Project director, project executive, or a senior responsible owner (SRO) are other terms for the project sponsor.
> - The **performing organization** is the organization whose staff is most directly involved in planning and executing the project. The project manager works for the performing organization.
> - **Core competencies** differentiate one organization from another. Such specializations determine eligibility for a project. A firm's employees' education and experience determine core competencies. A new project will either strengthen a business's existing core competencies or create new ones.
> - The **project scope** details all the work that must be done to complete a project. Negotiations between the client the performing organization determine the scope. The scope is essential for cost-estimating. Project problems usually arise when the scope is enlarged or diminished.

As mentioned in the introductory chapter, each of these specific definitions affects the others. For example, in smaller firms the project manager may have another duty on the project team. Moreover, there may not be a clear line between a potential client and other stakeholders interested in or paying for the project. A **stakeholder** is anyone whose interest in the project must be taken into account during project work. Stakeholders include project sponsors, members of the performing organization, anyone contracted outside of the organization to help perform tasks or help develop a product, and possibly potential customers or clients of the project or product itself. Stakeholders will likely be the first and last people a project manager interacts with over the life of a project. Stakeholders are the party most likely to have conflicts of interest during the project. Since stakeholders are not part of the project team, they may have an ulterior motive or undisclosed and compromising interest in the project. It is the first of many legal issues of which a project manager should be aware of from the get-go.

Stakeholders may drop in and out during the project's duration, but a project manager should keep track of everyone involved. **Stakeholder registers** identify, categorize, and evaluate stakeholders based on expertise and/or interest in specific areas of the project. The three steps of stakeholder analysis are: 1) identify stakeholders; 2) analyze potential impact; 3) assess stakeholders' likely reactions. To better understand stakeholders, grids are often employed. **Stakeholder grids** classify the power and influence of each stakeholder, from least to most in whatever category is being measured. The four grids are: the influence/impact grid, the power/influence grid, the power/interest grid, and the salience model. Salience refers to the power, legitimacy, and urgency of each stakeholder.

Stakeholders can provide expert judgment. For a firm considering a project in a new area, getting input from an experienced contact will provide insight on whether or not to initiate a new project.

Stakeholders should be invited to contribute to a discussion about performance on each major deliverable to improve future projects. Some deliverables, meanwhile, can be accomplished quickly while others linger and affect every other factor in a project. These muddled relationships cause much of the stress in project managing., which is why tracking and communicating with all stakeholders is so critical to a successful project.

The **three Is** of project management recognize this complexity. Every task one accomplishes in service of a project is integrated, interconnected, and interdependent. The slow flow of work in one area can profoundly disturb the sequencing of deliverables in another. Scope, quality, scheduling, budget, resources, and risks all make up some of the many obstacles a project manager faces. The **triple constraint** clarifies these competing needs and is displayed to the right. This model understands that a change to one of the big three areas of a project will impact the other. For example, if deadlines are aggressively pushed up closer to the present, the project would increase in price or have a more limited scope. The triple constraint can be depicted as a triangle or flow chart to emphasize these relationships:

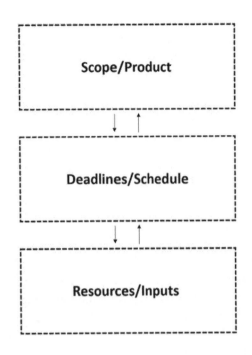

The "butterfly effect," modeled off of the concept that the flap of butterfly's wing could start a chain of events resulting in a hurricane, describes the effects one project area has on another. Such large unintended consequences are the norm in business. The history of economic crashes, disruption, and delays show that unexpected events are the norm. Understanding such complexity and having a big-picture view at a systems level is a necessity for project managing and a sign of business acumen. In fact, it is speculated that most projects fail. Being aware of the triple constraint, butterfly effect, and business norms will help project managers both prevent project failures and take a realistic view of a project's potential success.

The best way to anticipate a project's success is by completing a **work breakdown structure (WBS)**. A project manager cannot begin trying to resolve the triple constraint without having a WBS. A WBS is only created when a project is being seriously considered. A WBS defines the scope of the project, focuses on deliverables, shows how deliverables relate to one another, and is the basis for planning after a project is initiated. A WBS should include four levels: the project, deliverables, components, and work packages. The divisions of a WBS can be understood as enveloping circles:

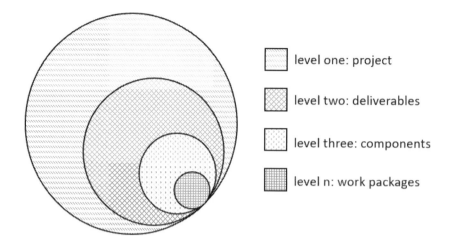

level one: project

level two: deliverables

level three: components

level n: work packages

The project itself is considered level one, the highest level of the project. During the initiation phase, most of the discussions will take place at the top level. Discussions about strategy and feasibility focus on the project as a whole. The WBS is more concerned about the specific deliverables, level two, that make up the project. Deliverables are in turn made up of components and work packages. The purpose of a WBS is to break the project down into work packages. A **work package** is the most detailed day-to-day description of project work. Analyzing each work package leads to early estimates of project duration and cost. Project components are made up of work packages and administrative work that must be done to facilitate work packages. A big part of project components is the project manager's supervision and organization of work packages.

A **project activity** can be any one of the three lower levels of a project. An activity is simply a specific portion of the project work that has separate parts that are similar. Different project activities often require different skillsets and personnel. Project managers decide which work packages will be included in an activity. **Decomposition** is the technique by which high level project requirements are broken down into distinct project activities. Decomposition is a necessary part of making a WBS. Often times, work packages are not obvious in the scope of a project's work. It is necessary for a project manager to decompose project deliverables into manageable work packages and then group similar work packages into activities. A **summary activity** or **hammock** looks to find related activities that can be grouped together to streamline later project work.

To sort through this complexity, a project manager must account for the type of firm in which they are employed. **Non-project-based firms**, like manufacturers who produce similar products over time, are not organized for project after project. Such businesses require an unfamiliar and upfront effort to put in place rules to complete a project. Non-projectized firms often have a **functional organizational structure**. A functional organization is divided into specialized departments, where each department works in one area instead of cooperating towards a larger goal.

The importance of a project within whatever type of organization will dictate resource allocation. Businesses with a **matrix** organization use many resources for multiple reasons at the same time and there is often competition for resources and unclear lines of communication. Team-members often have their functional manager to answer to and the project manager—a manager for the accountants and a manager for the project as a whole. Matrices are also used to show the relationship between resources, project team-members, and objectives. When diagrammed, matrices are an easy way to display data and show responsibility. In the simplified matrix on the following page, the functional manager is most responsible for the four team-members displayed. From the functional manager's perspective, the two lawyers and two writers are a team that produce shared outputs. A project manager, however, will group lawyer 2 and writer 2 for a project. A matrix for an entire project will have many more examples of sharing resources.

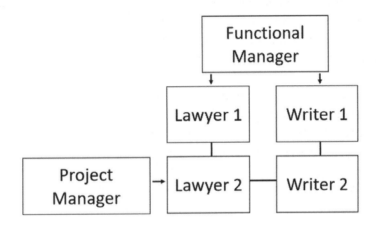

Project-based or **projectized** firms, like consultancies, usually have multiple ongoing projects at a time and an existing strategy for tackling projects. A project manager has more authority in a project-based firm than in a non-project based firm or a matrix organization. However, in projectized organizations, teams that work on a project are often ad hoc and lack a functional base that keeps them together after a project is complete. Regardless of the type of firm, knowing previous projects' failures and successes is an essential task to ensure a smooth initiation.

One of the biggest differences between project managing experiences is how much freedom a project manager has. Depending on the firm, higher management or ownership can be heavily involved in a project. The general rules are as follows: a functional firm gives a project manager very limited authority, a matrix organization gives a project manager a fair amount of authority, and a project-based firm gives a project manager nearly complete authority. As with everything having to do with projects, these rules of thumb depend on firm culture and personalities.

It is essential that, whatever the organization, clear chains of command are put in place once a project manager is selected. The vast majority of a project manager's time, up to ninety percent in some studies, is taken up by communication. Conflicting messages go out to the client and those authorizing the project worsens the triple constraint's messiness. Project managers, coordinators, and expeditors should all be on the same page once a firm assigns these roles. Everyone's distinct duties should be spelled out and the project manager is ultimately responsible for ensuring everyone stays in their assigned lane. Especially in bigger firms, it is essential to have a clear order of authority to decrease redundancy and have procedures for conflict resolution.

The type of firm also often determines how detailed the WBS will be. Depending on the type of project and the firm, **rolling wave planning** can be used to project future work. Rolling wave planning entails determining project plans over time, as details emerge, instead of at the beginning of a project. The project begins purposely with many unknowns that can only be defined by subsequent project work. A WBS in this scenario has many placeholders that will only be filled in as more information is gathered about the nature of project work.

Besides the issues that the type of firm and its methods of organization can cause, another wrinkle can arise from an overly-involved clients. Clients may ask for unfinished work and alter time constraints as they review unfinished work. Depending on the contractual situation, project managers are usually obliged to adapt to such changes. A way to mitigate such interruptions is to have the client assign a person on their side without a personal stake in the project to interface with the project manager. This allows for an unbiased and routinized relationship between the project manager and the client to develop. If that is not possible, figuring out the culture of the sponsor's firm and tailoring the level and detail of daily communication becomes necessary. The **project initiator**, whether inside or outside a project manager's firm, should be integrated into these early discussions on communication protocols. The initiator is the person that requested the project for a specific reason. The project sponsor may or may not be the same person.

Once an accounting of both the performing organization's and the potential client's modus operandi occurs and a project between the two is under consideration, a **feasibility study** is a typical next step. Looking at the required abilities to fulfill a project, a firm should decide whether this is a project they can or cannot manage. If it is determined feasible, the team in charge of the feasibility study should estimate the time, resources (new and existing), and individuals needed to complete the project. During this process, questions revolve around "could we?" and "should we?" thinking. Project stakeholders and initiators should be encouraged to detail what they hope the project to accomplish at a granular level. **Preliminary project scope statements** describe a project's objectives, the desired deliverables, and anticipated timelines before a firm commits to a project. All necessary deliverables required to complete the project are termed critical success factors in the preliminary project scope.

In answering these questions, a cost-benefit analysis can also complement feasibility studies by clarifying the outlook for a project and considering the consequences for the firm if the project fails. Discussions often center best- and worst-case scenarios and consider the **net present value (NPV)** of the firm's anticipated project expenditure. This value can be found if one is considering many projects by comparing the value of investing in one project over another by finding the profit difference. The net present value subtracts all the project's benefits and revenue from its cost.

If the NPV yields a positive amount, the project should result in a net gain in profitability or value to the company. If NPV is a negative amount, the project manager has determined the project will amount to a loss and should not recommend initiation unless there is an overriding strategic priority for initiating. If the first project in a new area results in a loss while the second project in that area would lead to a profit, that would provide a strategic reason to initiate. Nonetheless, the project manager should always recommend to not initiate if NPV is negative. It is up to higher management to take a gamble. Considering other demands on a firm's time and energy is also an important aspect of deciding feasibility. Project managers should ensure a project's anticipated deadlines are workable for the firm's existing calendar.

Every firm that initiates a project has a **business case** to justify the project. The business case takes the prospective project's costs, risks, and benefits into account to show that a project is worthwhile. A business case answers the "should we" questions referred to above. There are four typical justifications:

➤ A **benefits justification** initiates a project because the benefits exceed the cost and effort of a project. This is informed by the NPV analysis.
➤ A **compliance justification** initiates a project because a partner organization or a cooperation's headquarters needs it done.
➤ An **enabling justification** initiates a project because the project will have spillover benefits that will improve other operations.
➤ A **maintenance justification** initiates a project to provide an update or improvement to a past project.

In each of these efforts, a **scope statement** is a complementary and necessary step. It specifies, in writing, the results the project will produce and what the firm requires to do the work. The statement should specify the project team, costs, deadlines, and contingency plans for handling unexpected events. The statement will bind a project manager, the project manager's team and firm, the client, and stakeholders together to a shared vision. A legal contract could be attached to a scope statement to clarify the obligations of the firm and client. The **project charter** offers an understanding of the project's core goal and offers a preview of project participants and objectives. Compared to a project scope statement, a project charter is less detailed and is more high-level and strategic in its outlook. Once the project charter is signed, the project may begin. The project charter gives the project manager the authority to begin work.

During the initiation stage, a project manager should not try to map out specific, day-to-day activities months in advance. Committing to such a rigid timeline makes it more likely that activities or deadlines be will missed. Instead, the project manager should focus on the life of the project. **Project life cycle** is the sum of each project activity, from start to finish. The five phases of a project life cycle are: *initiation, planning, executing, controlling and monitoring,* and *closing*.

The traditional project life cycle moves along sequentially through the project phases. Most projects operate in this manner. For projects that must constantly alter plans, an iterative life cycle may be the preferred option. An **iterative project life cycle** is one in which the timelines and costs change as project work continues. The project scope is known, but the sequence of the work cannot be planned out. An iterative approach is often used when developing a product for the first time, when there are unknowns in the production process.

An **incremental project life cycle** prioritizes feedback. Work frequently pauses in order to gather stakeholder opinion. During these pauses, project work is tested and feedback is incorporated. There is a cycle of increments of project work, and pauses to evaluate that increment before the next increment of work can begin. An **adaptive project life cycle** is a combination of the iterative and incremental project life cycle models. Adaptive cycles must work in increments because the nature of the work is only found by doing it. The specifics of work at the end of the project are unknown at the start of the project. Adaptive project cycles depend on getting feedback after each increment of work, and then changing work techniques to meet that challenge.

A popular variation on the adaptive project life cycle is **agile project management**. Rather than depending on unknowns or pauses, the agile project embraces flexibility. Agile projects use sprints to work through the project—its increments. Sprints are short work cycles that focus on improving the project in a specific way. The work that occurs in each sprint determines the next sprint. Instead of working in the project from start to finish, agile managers work on parts in a more ad hoc manner as project needs arise. The point is to continuously improve the project and address the unexpected quickly. Agile management is especially common in software development projects in which issues in one area could have unintended consequences on another area of the project.

The **scrum framework** is another facet of iterative projects you need to know. Scrum is a rugby formation in which all of the team huddles together and tries to gain momentum in the game. The scrum framework dictates that what is known about the project at the start is negligible. Instead of having a plan, the team should be tightly focused on understanding the facts of a project and advancing together through conditions that may or may not change. In the scrum framework, incremental plans are tested for next steps but there are no large-scale, start-to-finish commitments. The scrum master plays the part of the project manager in this framework.

All forms of iterative project management, especially the agile variant, have limited to no need for a formal project manager. Project management is still needed, but the responsibilities are dispersed through the team instead of invested in one person. Teams are self-organized and event-dependent. This makes sense since an agile project responds to needs as they come up instead of following a detailed plan enforced by a manager.

Waterfall management however remains the most popular type of project management. In contrast to iterative forms, the waterfall methodology follows discrete phases to complete a project. Like the course of a waterfall, the project proceeds through tried and true phases. Sequentially, it goes through the process groups: initiation, planning, executing, controlling and monitoring, and closing. The organization of the following chapters assume a waterfall approach. But even an agile project will at some point have to deal with the budget. The budget sections that follow will be just as useful to an agile manager, although it may be useful during a different stage of project progress.

Thinking about the type of project life cycle being considered is a necessary part of the initiation phase. The life cycle type will determine the **integration management** style early on. Integration management is concerned with coordinating the competing needs of stakeholders and ensuring the tradeoffs of making a certain choice are understood. Integration in the initiation phase is tricky. A project manager may or may not be formally assigned until a project is initiated. Nonetheless, the future project manager is almost always involved in initiation thinking and will help in integrating project goals. For a firm of three or three thousand, there are best practices for a project manager to consider even before the planning stage begins.

Section 1.2: A project manager's mindset

The complexity presented above, from the triple constraint to managing opposing corporate cultures, is why project managing is in such high demand. Here are concepts that will help a project manager overcome whatever obstacle arises:

- ➤ A **milestone** or **project phase** is a crucial development in a project. Completing a large deliverable or realizing a deliverable that marks the end of the project are examples. Identifying project milestones will facilitate effective time-management. Milestones have specific start and end dates.
- ➤ **Line** or **functional managers** oversee a set of deliverables that requires specialized skillsets. Line managers assist project managers by allowing them to make informed decisions about resource allocation, scheduling, or costs.
- ➤ **Programs** are groups of related projects that are initiated simultaneously to achieve efficiencies. These projects require similar work. Grouping them together allows a firm to specialize.
- ➤ A **portfolio** groups many projects together for a common strategic, long-term reason rather than efficiency.
- ➤ An **application area** is a type of portfolio. Projects that share specific resources, business strategy, clients, or team-remembers are application areas. A project can fall into more than one application area.
- ➤ **Organizational project management** focuses on how programs and portfolios should be regulated to achieve a larger strategic goal.
- ➤ **Systems management** is a framework for understanding how part of a project affects the whole.

The best way to navigate through a project's many stages is to stick to a plan. The next chapter will walk you through proven and tested methods to formulate an excellent plan. Unfortunately, even in a project with relatively few crises, the stress of a project can make sticking to a plan difficult. Gathering information in the initiation stage on the project through conversations with stakeholders is key. Determining feasibility is far more than considering resources and cultures. It must include thinking through the stresses of a project for everyone on your team and considering their schedules.

In larger firms with many ongoing projects, a **project management office** handles staffing on multiple different projects and allows individuals working on multiple projects to prioritize their work. Smaller firms should be cognizant of the competing demands of multiple projects on team-members and clearly determine where their time should go in a similar fashion. There should be guidelines that shape how much of a workweek team-members devote to this or that project. Putting every decision in writing early on and sharing it with your team, no matter how small the project is, will payoff largely by the end of the process. Similar to a project management office, a **project steering group (PSG)** is for approving project work in one phase before allowing progress to the next stage. Important stakeholders, higher management, and project sponsors can all make up a PSG.

The extent to which this responsibility is the project manager's depends on how their organization is composed. An organization with a **balanced matrix** is one in which a project manager's authority is equal to the authority of each functional manager. A functional manager therefore has a lot of independence in their assigned work. In larger projects, the project manager is also given assistants with specific roles. A **project coordinator** assists the project manager, who is still responsible for the overall project. The project manager can delegate to the coordinator responsibility for various tasks. In large projects, a coordinator is responsible for keeping everyone on the same page. A **project expeditor's** primary responsibility is communication. A project expeditor works under the project manager, and cannot make any project-related decisions on their own authority.

No matter how simple or complex a task is, getting early buy-in from everyone involved will reduce stress later. Project managers should ensure no one on the team is caught off guard by a coming milestone. A **project champion** is someone higher than the project manager who strongly advocates for the project. A project champion can cajole team-members into reordering their priorities to expedite a project's completion and fight for more resources. These advocates should be included in the strategic vision of the project. In start-ups, the owner should champion each project and be on the same page as the project manager.

It is critical that all members involved in the project understand the scope strategy behind the project, beyond individual deliverables. When someone is brought onto the project, they should review the plan. Reviewing creates ownership and leads to another person thinking big-picture, who with a fresh eye may actually be more likely to catch mistakes. Even an outside subject-matter expert brought in for a single task should review prior work done. Duplication will be decreased and efficiency increased.

To keep everyone on the same page, objective to objective, a project manager should be succinct and avoid technical jargon and acronyms when possible. Employing **SMART Objectives**, or specific, measurable, aggressive, realistic, and timeline-driven objectives, will help keep the team cohesive and moving forward. There is a tradeoff between being objective-driven and adaptive to inevitable mistakes in a project. Objectives should therefore be ideals and goals, not promises.

Putting these safeguards in place enable a project manager to serve as the main motivator for team-members, as a project manager usually cannot on their own authority provide other incentives, such as pay raises or bonuses. The only thing a project manager can rely on is his or her ability to create a dynamic plan, and getting enough team buy-in that the plan survives intact to the project's end. A project manager should be singularly focused on putting their team-members in a position to excel. If a project manager can understand that, motivate team-members on cloudy days, and put the team's greater good over individual egos, a firm can confidently initiate a project.

1. A project manager is deciding whether to recommend a project to refurbish a number of buildings in a regional fast-food chain for whom she works. The NPV yields a negative value, but the owner and many stakeholders think the business needs behind the renovations are paramount. What should a project manager do?

    a. Recommend against the project for the time being
    b. Present the NPV alongside a separate report on the value of renovations, hear feedback, and then decide on recommendation
    c. Recommend for the project
    d. Let the board or any other entity decides whether to proceed or not

2. Which of the following the best captures the point when a project is complete?

    a. When all of the milestones are complete
    b. When the project team moves onto another project
    c. When all stakeholders' project needs are addressed
    d. When the project sponsor announces project completion

3. A firm is considering a project that requires the involvement of a copy editor who has been slow to communicate, and difficult to work with in past projects. During the initiation stage, how should a project manager expect to improve this copy editor's responsiveness?

    a. Plan on either decreasing or increasing the copy editor's future pay based on performance
    b. Clearly state the copy editor's duties ahead of time
    c. Get the copy editor's functional manager involved early on and ensure the manager understands the scope of the project
    d. Get the copy editor's functional manager involved early on and ensure the manager evaluates the copy editor's performance based on project performance

4. Which of the following qualities best captures the core skillset of a good project manager?

    a. Unfaltering self-reliance and self-motivation
    b. Quickly adapting plans to changing circumstances
    c. Initiating a detailed plan and sticking to it
    d. Facilitating interaction and productive communication

5. You have recently been promoted to project manager because the last project came in over budget and late. This new project requires you to update a retailer's displays in accordance with the latest studies in behavioral economics. The owner expects this change to quickly increase sales. Which of the following is not true in this situation?

    a. The trust in behavioral economics is an assumption on the part of the firm's leadership. The assumption may be wrong and represent group thinking about a fad. It needs to be investigated closely to ensure sales are likely to increase
    b. The objective is to do better than the last project manager
    c. The schedule is a constraint partly because the owner expects a quick result
    d. The budget is a constraint partly because the last project came in over budget

6. What is the difference between the project scope statement and the project charter?

    a. The project charter is for all stakeholders, while the project scope statement is only for team-members
    b. The project scope statement is specific and names the project manager, while the project charter is higher level and does not name the project manager
    c. The project scope statement describes the project in detail and specifies what project completion means, while the project charter is more general
    d. The project charter is completed before a firm commits to a project, and the project scope statement is written once a project has been initiated

7. How does rolling wave planning shape the work breakdown structure (WBS)?

    a. Milestones and targets are further detailed as the project evolves over time
    b. The first WBS has subproject placeholders instead of specific timelines for project completion
    c. A combination of some deliverables cannot be mapped out at the start of a project, and so the WBS is not completed at a single point
    d. All of the above

8. Two projects are under consideration by soda maker Pop Bottles. Project Y has a payback period of two years and Project Z a payback period of seven fiscal quarters. Project Y would generate a cash inflow of $30,000 per quarter and Project Z would generate $120,000 cash inflow per year. As a project manager, which project would you advise Pop Bottles to take ceteris paribus?

    a. Project Y
    b. Project Z
    c. They both have the same cash inflow per year, so both are equally advisable
    d. Not enough information to make a decision

9. Which of the following statements is true?

    a. Programs are groups of related deliverables
    b. Portfolios are groups related deliverables
    c. Hammocks group together overlapping project phases
    d. Project life cycles consist of overlapping project phases

10. You are taking over from a project manager moved to another department early on in the initiation phase. The departing project manager tells you that she was working on determining how the project would be executed and controlled. The project manager was working on which of the following?

    a. The project scope plan
    b. The project management plan
    c. The project charter
    d. The product scope statement

Section 1.4: Practice exam on initiating a project – answers

1. Answer: A. The NPV (net present value) includes profit and *value.* Ascertaining the value of a project, the provision in answer B, must be a part included in an NPV. If an NPV is negative, a project should not be recommended. In the initiation stage, a project manager is responsible for making a decision. The firm may go in a different direction regardless.

2. Answer: C. Stakeholders may ask follow-up questions or have related requests months after work on milestones ends. Often, contracts account for this dimension by requiring firms to answer emails on a product or service well past formal delivery and transfer.

3. Answer: D. A project manager (in that one role) cannot reward or determine pay for team-members, so answer A is incorrect. Even without work difficulties, answers B and C should be done. Following a difficult relationship, a project manager should be proactive and ensure managers tie incentives and disincentives to project performance.

4. Answer: D. A project manager spends an overwhelming amount of time during a project communicating. Keeping those channels open and facilitating a torrent of discussion is key to success. Self-reliance is an important quality for team-members, but a project manager depends on others doing the work. Planning without reliable and consistent communication will not lead to the desired outcome.

5. Answer: B. The objective is to increase sales via rearranging the retail displays. Every project includes assumptions that must be critically examined in the initiation stage. Time and budget are always constraints, no matter what happened during the last project.

6. Answer: C. Both project charter and scope are for all stakeholders, name the project manager, and are completed during the initiation stage. However, a project charter offers less detail and is easier to disseminate to stakeholders, while the timelines and work packages addressed in the project scope statement concern only those involved directly in the project. A work package is project work at the base level of the WBS.

7. Answer: D. Some projects have many subprojects and exact details that cannot be penciled in until a later date. These are dynamic and interdependent tasks. Therefore, the planning occurs in waves and the WBS is updated incrementally.

8. Answer: C. Ceteris paribus means "other things equal" in Latin. It is a common technique to make decisions between two similar options. In this case, project strategy and feasibility are assumed to be the same. Mathematically, the inflow is equal and there are similar payback periods.

9. Answer: D. A project's life cycle phases are initiation, planning, execution, monitoring and controlling, and closure. Monitoring and controlling overlap with execution.

10. Answer: B. The project management plan details the execution, monitoring, and controlling of a project. The project management plan has many subsidiary plans.

Chapter 2: Planning

Section 2.1: Teamwork planning

Surveying the many surprises and unexpected developments during his tenure, British Prime Minister Winston Churchill observed that "plans are of little importance, but planning is essential." More directly, boxer Mike Tyson stated that "everyone has a plan until they get punched in the mouth." What's the point of planning, then? Effective project planning should anticipate such shocks and have procedures to handle the unexpected. And every once in a while, a plan does work out more or less, saving time and effort. The most important plans deal with the human resources—stakeholders and the project team—that the project manager must manage.

The project team is one subset of the total amount of people that must be involved in a project. However, the number of stakeholders often outnumber the amount of people directly working on a project, and stakeholder involvement should be a constant during every stage of a project. Getting stakeholders to support the project's development is a critical element of success.

Effective communication, or customizing messages for each audience involved in a project, is needed whether interacting with a rarely heard stakeholder or a vital subject matter expert. Emails, memos, or reports should include only necessary information and be to the point. Actions verbs, active voice (as opposed to passive voice), and using the fewest and shortest words possible are good tips. The difference between effective and **efficient communication** is an efficient communicator provides information in a timely fashion. Efficiency mandates that a project manager reminds their team well before deadlines and is aware of potential issues long before they can arise. As mentioned in chapter one, since communication takes up the vast majority of a project manager's time, there should be a good deal of self-reflection about communication patterns.

Understanding the **lines of communication** within a project is a best practice. Lines of communication refer to the total number of possible communications between each member of a project team. Thinking about the total network of communication, project managers talk about **nodes** that create and determine lines of communications. Nodes are individuals, departments, or some other entity that communicates with the project manager during the project. In the diagram on the following page, there are six nodes and fifteen possible lines of communications between them:

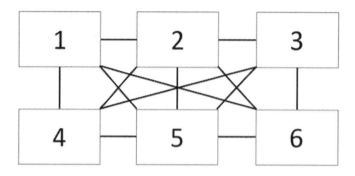

Rather than drawing it out, one can determine the total lines of communication with the following equation:

$$lines\ of\ communication = \frac{n(n-1)}{2}$$

Plugging in six nodes leads to the same answer:

$$\frac{6(6-1)}{2} = \frac{6(5)}{2} = \frac{30}{2} = 15$$

Controlling lines of communication and placing yourself in the center of all communications is a necessary feature of effective project managing. A project manager that loses control of lines of communication loses control of the project.

A **project management plan** is the collection of many, more specialized plans, from regulating communication to closing a project and everything in-between. All of the inputs that make up the larger project management plan are known as **subsidiary plans**. We have already encountered a number of these above (scope management, risk management, communication plans, and the project charter).

Every plan should contain objectives and these objectives should be **SMART**: specific, measurable, achievable, realistic, and timeline-driven. Being smart about objectives and sorting them appropriately will mitigate the chances of the project's scope overwhelming the team. The three techniques covered in chapter one to breakdown interdependent or complex, multi-step objectives were: decomposition, rolling wave planning, and expert judgment. When crafting objectives, a consistent level of accuracy should be used across the entire project management plan. Whether an hour or month is being rounded, the same unit of measurement should be used for each objective that is tied to a deliverable. All of these principles help in preparing the **activity list**, which covers the scope of work for each scheduled activity.

Activity lists should be a collaborative process that goes through multiple drafts with input by multiple prospective project team-members. Assumptions are often corrected when listing activities. **Assumption analysis** is an important step in the planning process. The analysis finds all the implicit thinking done about the project and checks the premises for accuracy and completeness. It is important to bring the entire project team together so planners do not overestimate capabilities and can synchronize calendars around the project. For example, an engineer that is slated to work on the project will be much better at checking mathematical assumptions than a team-member lacking that expertise.

The **human resource management plan** is an essential subsidiary plan, as it defines roles and assignment-owners within the team. It also sets the hierarchical order within a project to establish lines of communication and reporting requirements. A project manager does not have to reinvent the wheel when it comes to managing human resources. Beyond the project team, stakeholders also need a plan. The number of stakeholders and the weight of stakeholder input decides whether a project requires a **stakeholder management plan.**

Consider **enterprise environmental factors (EEFs)** when making subsidiary plans. A firm's culture, history of successfully managing projects, business partners, assets, and knowhow, such as subject-matter expertise, all qualify as internal EEFs. Environmental factors can also be external and make the project more difficult to complete, such as location, distance, market changes, and government regulatory conditions that can affect the firm, but not necessarily shape the project directly. Legal and labor requirements should be considered in the entire project plan and especially with schedule planning. Calling your attention to context, EEFs try to ensure a plan considers the many possible ways a project can play out.

After considering EEFs, moving on to consider **organizational factors** will allow a project manager to decide which departments and/or teams within a firm will be responsible for what during the project. Organizational factors include the makeup of your firm, its hierarchy, risk management, financial restrictions, safety standards, and rules governing how departments interact with one another. The human resource management plan should take advantage of enterprise environmental and organizational factors. Policies on reporting requirements, overtime pay, communication protocols, safety requirements, and anything else that governs firm policies should be spelled out. Personnel policies, regarding rules for vacation time, daily breaks, and hiring, firing, and assigning tasks to team-members, should inform the human resource management plan. The resources made available to a project manager during a project, human or not, are often conditional and can change. However, the rules governing those resources usually last for many projects.

Interpersonal factors can prove to be critically important. Experiences, skills, team dynamics, personality types, and cultural sensitivities should all be accounted for and openly discussed to ensure team cohesiveness. A project manager over time should realize which combinations work well together. To assist in appraising a new combination of team-members, a project manager should seek out advice on team-members' tendencies. Technical factors, regarding the specialized skills a project requires like architectural design, are more straightforward. Nevertheless, a project manager should be careful to avoid incompetent experts or experts needlessly extending the time to finish a task. For many organizations, information technology tasks see such extensions. To maximize efficiency, a project manager should research the typical time to complete an expert's work in analogous projects. There are also location and logistical factors to consider, defining how a project team will meet, coordinate work, and which standards will be used, in-person or virtually.

To balance these many constraints, many project managers use a **responsibility assignment matrix (RAM)**. A RAM ties deliverables or activities to individuals on the project team to provide clear lines of responsibility. **RACI charts** are a useful and popular type of RAM. RACI is an acronym detailing that chart's usefulness:

➢ Responsible: who does the work to complete an activity
➢ Accountable: who reviews the work done
➢ Consulted: who is the individual responsible for the work rely on for guidance
➢ Informed: who is notified when an activity is completed

RACI charts are often quite large and encompass the entire project team, from a project manager's assistant to a subject matter expert (SME) used only once. Here is a simplified RACI matrix:

	SME 1	SME 2	SME 3
Activity 1	RA	I	C
Activity 2	C	R	AI
Activity 3	R	A	C
Activity 4	A	CI	R

A **linear responsibility chart (LRC)**, is a type of RAM which ties deliverables or activities to individuals on the project team to provide clear lines of responsibilities. It is a streamlined RACI chart.

Section 2.2: Budget planning

Stakeholders are naturally most concerned about the budget, so it should be a project manager's second highest planning priority after the team. Stakeholders may have different opinions about the budget, and some departments and individuals involved in the project may not know how their expenditure affects other budget lines. It is your job to ensure everyone is on the same page and shares a whole-project perspective when it comes to itemization.

As with all plans, the budget likely will not pan out just as it was drawn up. There will likely be excesses, as coming in under budget entails a plan working out nearly flawlessly or there were unexpectedly positive negotiations about price. Two concepts are fundamental to budget management. First, **opportunity cost** is the loss incurred when one path is taken over another. It is the best path not taken; the alternative, optimal use of the time spent on a task. If a project-oriented firm chooses one project that takes three months, their opportunity cost could be another project that would have lasted three months and imparted different benefits. Every project has a large opportunity cost, as a firm could have chosen a totally different project instead. For a budget, opportunity cost compels a project manager to think about how else funds could have been spent. **Economic profit**, as opposed to accounting profit, factors in opportunity cost to consider revenue a firm could have acquired during the time spent executing a project. As with staffing a team, a consistent level of accuracy (similar rounding, same metrics, etc.) is essential for a robust budget plan.

The second concept fundamental to budget management is the **cost management plan**. A cost management plan is the main output in budget planning, a subsidiary plan. It details how the project's expenses will be determined, monitored, and controlled. Using the work breakdown structure's (WBS) list of activities and work packages, the estimated costs process is a key, early step. The estimated total budget is deduced from these smaller estimates. Critically evaluating the sequence and duration of every task, structuring fund releases around milestones, and figuring out the payment processes that will be used will make the plan more effective.

At this early stage, the **estimate at completion (EAC)**, which is the total project cost estimate as well as estimates by deliverable, is provided. These are all static numbers determined before a project's execution phase begins. The **estimate to complete (ETC)**, meanwhile, is a dynamic number that is calculated when a project is ongoing. An ETC should decrease as the project's end nears, unless there are drastic misestimates that lead to high costs as the project matures. Using a consistent format or online program to handle and oversee the budget from the first plan to last payment produces better budgets.

There are many strategies to estimate costs. **Bottom-up cost estimating** uses the WBS to map out a project's cost, dollar-by-dollar. Earlier in the initiation stage, **top-down cost estimating** is less accurate, but quicker. It looks at only the top levels of the WBS and estimates by milestone. **Analogous cost estimating** examines the budgets of similar completed projects within the firm or accessible budgets (for example, government contractors) from outside the firm to estimate costs. Analogous estimating is a form of top-down estimating. If a firm has a history of similar projects, a top-down estimate is far more likely to be accurate. Another top-down estimating strategy does not depend on the WBS: **parametric cost estimating** uses mathematical formulas to estimate costs of a deliverable by using the relationships between variables like per unit cost. A manufacturing company might use parametric estimating to approximate its variable costs on a per-item basis by multiplying quantity by rate and per-unit cost and factoring in gained efficiencies. Similarly mathematical, **regression analysis cost estimating** looks for statistical trends between the costs of similar deliverables to estimate costs at varying scales.

Estimates are often thrown off in larger projects. Construction, events, and similar projects even have "acts of god" clauses to protect the executing firm from large, unexpected events. Such disastrous occurrences can cancel or set back a project by months. A project manager should never forget about such worst-case scenarios and ensure legal, contractual protections are in place when necessary. **Three-point cost estimating** turns this necessity into a virtue. Its cost estimate is the average of the likely, worst-case, and best-case estimates. A useful way to plan for "worst" or "best" cases could be via slowest and fastest timelines for a project's completion.

The **budgeted cost of work scheduled (BCWS)** provides totals for all types of specialized work over the entire scope of a project. **Cost of quality (COQ)** refers to the expenditure required to achieve a specific standard in the project. Estimates for varying quality are integral parts of three-point estimates' best-to-worse case cost scenarios.

There are six theories, developed by seven individuals, which add more nuance to a project manager's understanding of COQ. You may be asked to identify the theorist or theory. For the purposes of the exam, one should know the theorist and the theories' titles and buzzwords:

- ➤ **Philip Crosby and absolutes of quality management.** Crosby is known for championing "zero defects" and trying to prevent redoing tasks in projects with large supply chains.
- ➤ **W. Edwards Deming and the plan-do-study-act (PDSA) cycle.** Deming thought of quality as chiefly the responsibility of project managers. The PDSA tests a change and widely shares the results so a method can be improved. The PDSA is action-oriented and involves experimentation.
- ➤ **Armand Feigenbaum and total quality management (TQM).** Taking a systematic view of an organization that is cohesive, TQM encourages a constant mindset of continuous improvement to provide increasingly higher quality products.
- ➤ **Joseph Juran and fitness for use.** Juran is most concerned with design, safety, and conformance. He developed a ten-step program to implement what he dubbed "quality by design."
- ➤ **Walter Shewhart and the plan-do-check-act (PDCA) cycle.** The PDCA also focuses on continuous improvement. Shewhart developed a four-step process to improve processes, as well as products.

> ➤ **Bill Smith and Mikel J Harry and six sigma (6σ) measurements.** Smith and Harry focus on output and believe an optimum number of allowable defects rounds out to 3.4 per 1,000,000.

Additionally, the Japanese word **kaizen**, meaning "continuous improvement" and the need to restlessly watch for opportunities to improve, is also associated with COQ thinking. **Grade quality** is another dimension. Grade quality, often used in engineering projects, differentiates quality not by cost, but by technical requirements and capabilities. In the case of a bridge, for example, grade refers to how closely various components meet or exceed load-bearing requirements. High-grade materials or equipment does more than or exceeds the minimum. Standards for grade in industrial projects are often set by American National Standards Institute (ANSI), or its partner organization, the International Standards Organization (ISO).

Quality and grade are important to avoid wasted funds. **Internal failure costs** occur when a product or process is still in development, but deemed by the project team to be inferior and in need of a rework. It often throws the budget and schedule off, but is better than the alternative. **External failure costs** occur when the customer, project sponsors, or regulatory agency outside the firm find the product or process to be inferior. The earlier this occurs, the better. If it occurs towards a project's close, final project costs may be double the original estimate. Inviting outside sponsors and stakeholders to review the project in earlier stages can make external failure costs more manageable. Moreover, and regardless of the level of quality or grade chosen, a **process improvement plan** hones in on inefficiencies, can help mitigate the chance of incurring failure costs, and can help a project meet cost estimates.

One of the most long-lasting products of all cost estimations is the **budget baseline,** which is the original, final cost estimate before project work begins. There are also baselines for the scope and schedule. Budget baselines on a graph have an S-shape, representing the initial lower amount of spending for a project, the cash need accelerating, and the final slowdown as the project ends. **Cost variance** is calculated at the end of a project and is the difference between the budget baseline and the final cost. **Control thresholds** are a critical component in limiting variance. They define the amount of slack a project manager has in allocating additional funds before seeking out more funds or reducing the scope.

A sunk cost, or funds that have already been spent, is not affected by estimates and not accounted for in control thresholds. A sunk cost can also be money spent to purchase capital for a past project that will again be used in a new project. The **fully burdened rate** is the total amount needed to sustain all of a project team's resources, from utility bills to every salary. Project managers track the burden rate and estimate the future consequences of hiring more help in the present. **Allowable payback time** complements these amounts by specifying the deadlines when payments relevant to project must be delivered.

For businesses conducting an international project or one with a long timeline, the foreign exchange rate and **discount rate** also shape budgetary decisions. The discount rate attempts to estimate the future value of today's currency, per changes in inflation and currency exchanges.

Many budgetary items interface with scheduling. **Resource calendars** allow a project manager to target costs of a particular category of resource over various time scales. Resource calendars detail when project resources will be available. The **resource breakdown structure (RBS)** divides resources up into these categories and is an essential tool in cost estimates. Resources common across many projects include salaries, rent, subject matter expert fees, and online utility costs.

In today's service-dominated economy, salaries and personal fees will likely comprise a large chunk of a budget and be the project's primary resources. The **performance measurement plan** enables the tracking of work done in achieving goals and the evaluation of those on the project team. Keying into performance shortcomings will make failure costs more avoidable, and make it more likely costs stay beneath the baseline.

Realizing there will likely be imperfections in the project plan, the **risk planning processes** completes the budget planning process. Risk planning entails putting procedures in place to deal with late or inefficient work or anything else that negatively impacts the project. You cannot plan for every eventuality, but you can plan the steps to take when things fall short. **Contingency reserves** are spelled out during risk planning and are built into the cost baseline. Contingency reserves provide slack for work most likely to go wrong and allow for minor setbacks to be easily addressed and funded. **Management reserves** are reserves for unforeseen obstacles and are therefore not funded in the cost baseline. Management reserves are funded from a different account, and used if the project scope changes, or expected resource costs change.

Section 2.3: Schedule planning

The **schedule management plan** pulls team and budget planning together and defines the flow that will structure the project. It defines how the schedule will be developed, monitored, and, if needed, altered. Managing the schedule is a dynamic process. As seen in the risk management process and contingency reserves, a task is likely to face unexpected challenges and the timing of subsequent tasks will be thrown off. A **master schedule** sets due dates for all deliverables and highlights milestones around which deadlines cluster. A master schedule should be widely available and updated with new information. What the responsibility assignment matrix (RAM) does for teamwork, the master schedule does for the schedule. It summarizes all schedule information. Organization charts often show interrelationships between team-members as determined by deadlines and the need to work together during the same time window.

Think of a schedule as a type of system, where each individual deadline is part of a whole and dependent on other dates. Schedule planning therefore requires even better estimates than the team or budget. More people can be hired and more money spent, whereas time is a far more finite resource. In fact, more project plans go awry because of delays than any other factor. To nail down more accurate estimates, schedule-planners often employ the **program evaluation and review technique (PERT)**. A PERT tries to determine accurate timelines before project work begins. In its simplest form, a PERT weighted equation is used to find the mean estimate using optimistic (O), pessimistic (P), and most likely timelines (M from mean) for the entire project or a specific deliverable:

$$PERT = \frac{O + 4M + P}{6}$$

PERT is most effective in figuring out duration. The **precedence diagram method (PDM)** is a more precise way to map out the schedule's interrelationships. The PDM uses boxed nodes that represent activities and arrows to represent the sequence and dependencies between each activity. There is often a specific order in which activities must occur, and a precedence diagram illustrates this. In the diagram below, the left edge of each box represents an activity's beginning while the right edge represents the same activity's end. The PDM utilizes four dependency relationships between tasks. Here are the four definitions and their illustrations, ordered from most to least common:

- **Finish to start (FS)** describes a relationship where one activity cannot start until another one finishes. A second (or "successor") activity only begins when the work for a first ("predecessor") activity is complete.

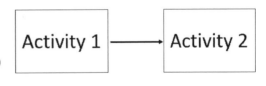

- **Start to Start (SS)** describes a relationship where one activity cannot start until another one starts. A second (or "successor") activity only begins a when a ("predecessor") activity also begins. Often, these activities begin concurrently.

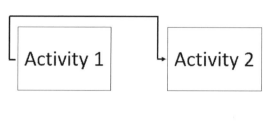

- **Start to finish (SF)** describes a relationship where one activity cannot finish until another one starts. When a second (or "successor") activity is complete a ("predecessor") can begin.

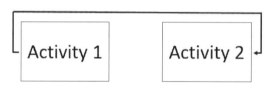

- **Finish to finish (FF)** describes a relationship where one activity cannot finish until another one finishes. A second (or "successor") activity is only complete when ("predecessor") activity is also complete. Often, these activities end concurrently.

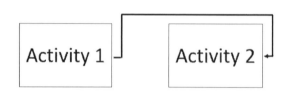

With schedule planning, these directional dependency relationships must be known to sequence to a project appropriately. The difference between SF and FS has to do with just such a dependency relationship. SF relationships are very rare, occur mostly in construction projects, and arise when there are delays or newly discovered requirements. **Mandatory relationships** express such sequencing constraints. Certain activities can only be done in a certain order. Mandatory relationships are also termed **hard logic**. Especially within large, formalized project teams, the hard logic in the ordering of tasks must be identified to stay organized. All four PDM relationships are mandatory.

Discretionary dependencies refer to a sequence of tasks not dictated by the nature of the work, but rather agreed to within the project team. A set of tasks that can be done in a varying orders without delaying the project is an example of a discretionary dependency. A firm's entrenched practices or an industry's standards lead to discretionary practices. Discretionary dependencies are also termed **soft** or **preferred logic**.

Outside of project work, other factors may interfere with workflow and require a change to the schedule. An **external dependency** is an event that occurs outside the project scope that can change the project schedule. For instance, if a law changes educational services for a disabled student, an education-technology project would have to change the order of its tasks to reflect this new law. An **internal dependency** is an event that occurs within the project team, but not tied to the execution of tasks, that can change the project schedule. An example of a delay to an internal dependency would be a project team member having to leave their job for family reasons, resulting in postponed work for the tasks they handled. It occurred within the project team but is still not directly related to the project work.

Combining PERT's emphasis on duration and PDM's emphasis on activity relationships, **schedule network analysis** is a necessary step to plan out the project schedule. The analysis produces a network diagram (also known as an arrow diagram) in which the duration and dependencies can be viewed:

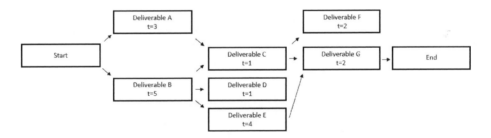

Per the level of accuracy requirements discussed above, the time (t) unit should be the same for each deliverable. In this example, time is measured in weeks. The dependency and duration information displayed in the diagram can also be displayed in a table. The arrows represent options. From Deliverable B, you can go to C, D, or E or all three.

The network diagram yields a good deal of information critical in planning out a project's schedule via two methods. The **forward pass** scheduling method calculates early start and end dates, moving forward from the starting node going through each deliverable. A forward pass method finds the longest path in the network diagram. The **backward pass** scheduling method begins at the end date to determine the longest delay of an activity without extending the project's entire timeline. Forward pass adds time by looking at when deliverables start, while backward pass subtracts by looking at when deliverables end.

The **critical path** is the sequence of deliverables that takes the most time to finish the project (to get to "end" node above) by using forward pass methodology. In this example, the critical path sequence is B-E-G for a total of eight weeks. All other options to get directly to "end" take less time. Project activities that can be skipped in a critical path, such as deliverable A above, is **sub-critical.**

Float or **free slack** refers to the latest possible delay a deliverable can be started without delaying the project. Deliverable C can be begun three weeks after work on deliverable E starts without causing delays to dependent work. The **latest finish date** uses the backward pass to find the furthest off date in which the project work can be completed while still in line with the relationship and dependencies between project activities.

If there is a delay, **fast-tracking** is a technique project managers employ when they can. Fast-tracking brings up a deliverable and begins work on it, while other work is going on. In the example above, deliverable F can be fast-tracked once work with C is completed, even if work on E is still ongoing. Rather than a delay, sometimes a project task is allocated too much time, or is given extra time in the likelihood unexpected obstacles pop up. **Feeding buffers** allow for certain activities that feed into the critical path more time than needed to be completed, so to ensure they do not delay many deliverables depending on them.

The remaining component to consider when schedule-planning is resources. **Critical chain method** changes the project schedule to account for scarce resources. Due to its focus on resources, critical chain method by necessity uses feeding buffers to ensure a resource can be spread around within the project team and be available when needed. **Resource leveling** (also called resource-based methodology) occurs when multiple activities need the same resource at the same time, or recourse is available only at certain times, meaning that a project manager must look to ensure resources are allocated by priority within the project schedule.

Resource smoothing changes activities' float times to increase resource availability. Deciding whether to change the critical path is a decision project managers must make when either leveling or smoothing out resource use. Managing resources within the schedule requires balancing project needs and team-members' time. Fortunately, there are ways to get back on track if resource issues are causing delays to the project. **Resource crashing** adds additional resources to expedite the project's critical path. The **stage-gate process** is useful when there are risks about doing many activities at the same time. "Gates" are set up at strategic points to monitor work. Before work can continue, the activity's progress must satisfy a checklist before the project can pass through the gateway to the next set of activities. Anytime a resource is only used once and is dependent on other activities, **reverse resource allocation** should be used to schedule the resource's use. A project manager works in reverse because task dependencies can be more easily checked from the end date.

Section 2.4: Diagramming to assist planning

Graphics are a mainstay of project management best standards. They assist all phases of the project, but appear most often in the planning stage. Some planners, especially in consultancies, use graphics too much. There should be an ideal middle ground where diagrams do not overload team-members, but are still convenient ways to communicate detailed information. Here is a primer on nonobvious diagrams you are likely to encounter, ordered from most to least common.

Histograms are bar graphs and a statistical favorite. The horizontal axis below is an occurrence category and the vertical axis is that category's frequency. In the example below, there were eight occurrences of numbers between 1 and 2.5 in a customer satisfaction survey.

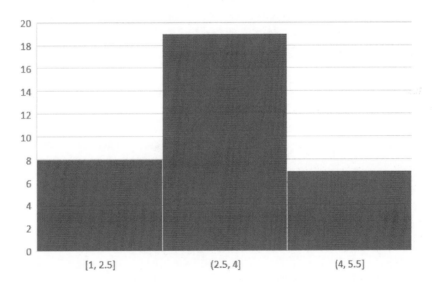

A **cause and effect diagram's** purpose is to find the origins of project problems. In the idealized example below, there are four large causes of a project delay, and each display two possible triggers. It can have far more branches than two and is usually used to display issues at the work-package level, rather than a macro issue such as a delay. This type of diagram is also referred to as a fishbone or Ishikawa diagram:

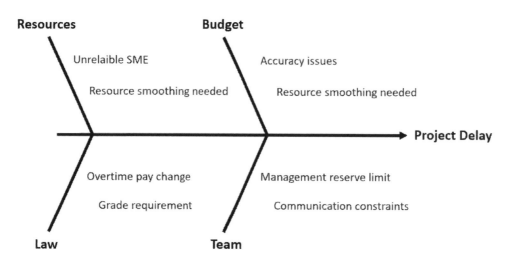

A **Gantt chart** has a vertical axis representing time, and horizontal bars reflecting how long a project task is projected to take. In the example below, Deliverable 1 is composed of four activities and its entire length is made up of the duration of the four activities. Activity C, for instance, lasts six weeks, from week four to week ten.

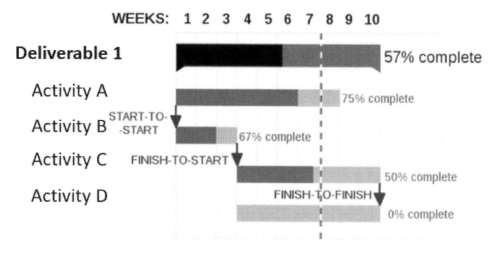

Pareto diagrams have two distinct vertical axes. On the left vertical axis are frequency measurements representing how often the phenomena on the horizontal axis appeared. On the right vertical axis are cumulative percent measurements of each occurrence from left to right. As you can see, the nearly 60 incidences of late arrivals caused by traffic account for about 34% of all excused; childcares' nearly 45 incidents accounted for 27%; and together they both accounted for 61% of all late arrivals. The size of each bar and slope of the cumulative percent line help a team see which trends matter the most.

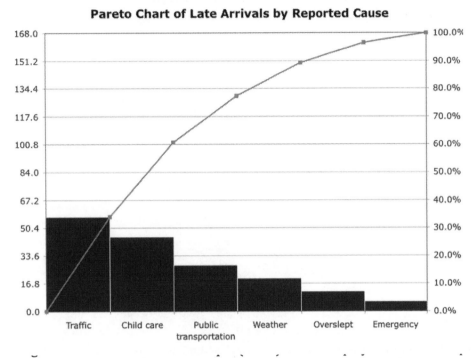

between project activities. Nodes are deliverables in projects and arrows represent events, as in network diagrams. Arrows are not drawn at scale to represent time, which in the example below is measured in months. To determine the time between project tasks, PERT employs probability to estimate the total duration of the project.

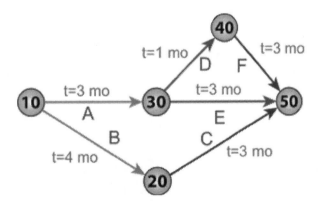

Control charts are mostly drawn up for industrial or manufacturing projects in which minor physical defects could set the project back. In the example below, each dot represents a sample from large sets to individual items, depending on the preference. The dots represent the average quality of the sample based on a grade or other metric. In accordance with grade requirements and the client's needs, the project manager should set the limits that are tolerated and investigate samples that are close to or exceed each limit. Although not reflected in the example below, standard deviation often provides the upper and lower controls. For the exam, one should know that in a **normal distribution**, one standard deviation includes everything 34% above and below the mean (68% total), while two standard deviations include everything within 47.2% of the same mean (95% of total).

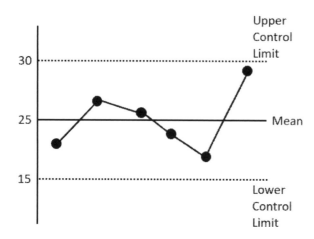

The **graphic evaluation and review technique (GERT) chart** uses network analysis for complex projects, often in engineering, and allows feedback loops, repetitive steps, alternative paths, and probabilistic and conditional branching. In the example below, circles are nodes (deliverables), the number on the right half-circle represents sequence, the numbers in the left quadrant-circles represent required loops, and the three numbers in-between arrows represent probabilities for each of the three outcomes and must add up to 1 (100%). Often, GERT charts have dozens of nodes, organized both horizontally and vertically, but the relationship between any two resembles the simplified relationship below. Running mathematical Monte Carlo simulations provides the numbers in a GERT chart. These simulations randomly sample outcomes using random methods to try to quantify the totality of possible outcomes. Since the project manager test is for the generic project manager, this description is all that is required for GERT charts.

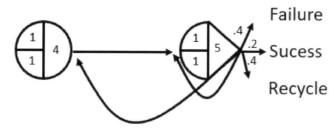

Section 2.5: Practice exam questions on planning

1. Which of the following illustrates what Armand Feigenbaum contributed to cost of quality (COQ) understanding?

   a. Total quality management
   b. Plan-do-check-act cycle
   c. Six sigma
   d. Continuous improvement

2. What are the similarities and differences between hard and discretionary logic?

   a. Both are used to define project dependencies. Hard logic orders tasks in a required order, while discretionary logic handles isolated and optional tasks in any preferred order.
   b. Both are used to define project dependencies. Hard logic orders tasks in a required order, while discretionary logic allows tasks without dependencies to be completed in any order.
   c. Both are used to define the project scope's sequence. Hard logic orders tasks in a required order, while discretionary logic allows tasks without dependencies to be completed in any order.
   d. Both are used to define the project scope's sequence. Hard logic orders tasks in a required order, while discretionary logic handles isolated tasks in any preferred order.

3. During an information technology project, problems arise about quality and grade. In a meeting, a subject-matter expert makes the following statement:

> *"From the start, it was clear that if we paid less and lived with lower grade equipment that we would pay a greater price later. We went ahead and bought lower grade equipment to make this product, and as anyone would expect, our product is now low quality."*

If she is, how is the subject-matter expert incorrect?

a. Grade and quality measure the same thing but using different standards
b. Grade and quality measure different things using the same standards
c. Grade and quality are not necessarily connected
d. The subject-matter expert is correct about the relationship between grade and quality

4. What are the benefits of a RACI chart?

a. It allows a project manager to analyze and control lines of communication
b. It displays cause and effect between various project activities
c. It projects the roles of the project team in various project activities
d. It illustrates the sequence and dependencies between various project activities

5. A project manager has just finished reviewing the activity list schedule by using analogous and three-point estimating. The project manager was engaged in which of the following processes?

a. Developing the schedule
b. Defining activities
c. Sequencing activities
d. Estimating activity durations

6. What is the relationship between organizational assets and enterprise environmental factors?

    a. Organizational assets are a subset of enterprise environmental factors. The latter includes all internal and external environmental factors that can shape a project. Organizational assets are just one type of internal factor.
    b. Enterprise environmental factors are a subset of organizational assets. The latter includes all processes and experiences that can shape projects. Enterprise environmental factors are just one type of asset.
    c. The two are independent of each other. Organizational assets are internal to the project team, while enterprise environmental factors are external to the project team.
    d. The two are independent of each other. Organizational assets have to do with processes and experiences, while environmental factors have to do with firm culture and resources.

7. A project manager is overseeing project activities done in parallel and is nervous about the risks of doing too much at the same time. The project team planned that there will be certain pause points to decide if activities should be done in succession or concurrently. Reviewing activities in this manner is known as which of the following?

    a. Rolling Wave Planning
    b. Stage-gate process
    c. Decomposition
    d. Expert judgment

8. Which of the following is an example of parametric cost estimating?

    a. A project manager compares costs from previous projects to extrapolate costs for a current project
    b. The cost of every single work package is calculated in detail and the total project cost is the sum of these smaller costs
    c. A project manager calculates cost by looking at most likely, pessimistic, and optimistic costs and works with the average of the three
    d. None of the above

9. A project manager leads a team of thirty people. Some individuals communicate daily, and others are silent for days at a time. More often than not, the project manager is carbon-copied on emails. How many lines of communication exist in this project?

    a. 465
    b. 930
    c. 870
    d. 435

10. There are three activities: A, B, and C. Activity C is quality control of activity B. Activity A is the work needed to prepare for the quality control executed in activity C. If A and B have the same duration, what is the relationship between activities A and B?

    a. Finish the start (FS)
    b. It is a case of soft logic
    c. Start to start (SS)
    d. Start to finish (SF)

Section 2.6: Practice exam questions on planning – answers

1. Answer: A. As defined in the chapter, COQ "refers to the expenditure required to achieve a specific standard in the project. Estimates for varying quality are usually integral parts of three-point estimate's best to worse case cost scenarios." It is mostly used in projects with a physical output. Total quality management (TQM) is a project quality theory developed by Armand Feigenbaum. Taking a systematic view of an organization that is cohesive, TQM encourages a constant mindset for continuous improvement to provide increasingly higher quality products.

2. Answer: B. Discretionary logic, also known as soft logic, allows the project manager to set an order between two tasks because there is not a mandatory dependency between those two tasks. Yet soft logic deals with related, not isolated tasks. Hard logic sets a required order between two tasks because there is a mandatory dependency between the two. Both logics are most related to dependencies. Since discretionary logic can be changed without any consequences, it is not crucial when setting up the project scope's sequence.

3. Answer: C. The subject-matter expert sees a clear connection between grade and quality, which is not necessarily the case. Quality has to do with condition (such as durability). Grade has to do with features, which are a choice. A project cannot begin with equipment with an incorrect grade. Lower-quality equipment would make a worse product, not necessarily lower-grade. The subject-matter expert should have said "lower-quality" because a higher grade would not have addressed his concerns.

4. Answer: C. RACI is an acronym that represents the process of determining who is responsible, accountable, consulted, and informed per each project activity. Answer A is done through a basic network diagram and the lines of communication equation. Answer B evokes cause and effect diagrams (also known as fishbone and Ishikawa diagrams). Answer D describes the precedence diagram method.

5. Answer: D. This is a clear case of estimating durations, as that is the point of analogous and three-point estimating. Estimating activity durations can use more methods than those two. Developing the schedule looks at total time estimates, defining activities is not concerned with scheduling but instead lists activities, and sequencing looks at network relationships.

6. Answer: D. Organizational assets are policies, processes, and guidelines that assist in project execution, as well as experience gained from past projects like contacts made with influential stakeholders. Enterprise environmental factors are either internal to the project team, such as culture and available resources, or external, such as consumer demand and the law. Notice that answer C is incorrect because enterprise environmental factors are often found within the project team.

7. Answer: B. The stage-gate process is used when there are risks about doing many activities at the same time. Gates are set up at strategic points to monitor work. Before work can continue, the activity's progress must satisfy a checklist before it can pass through the gateway.

8. Answer: D. Parametric estimates look at historical data and multiple variables to figure out the cost per unit and then multiply that by the amounts of units needed. If a firm is responsible for installing windows, the number of windows multiplied by the cost for installing one window would be a simplistic form of parametric estimating. Answer A is analogous estimating and answer C is three point estimating. Answer B illustrates bottom-up estimating. Bottom-up is the most accurate and time-consuming form of estimating. It comes from the work breakdown structure (WBS) to plan out spending dollar-by-dollar. However, bottom-up estimates require a lot of information and are usually not able to be done in advance of project work being carried out. Analogous estimating is the least accurate of the four and three-point estimating is more accurate than parametric estimating.

9. Answer: A. All the information about who talks and who does not is irrelevant. The project manager is part of the team meaning, there are thirty-one nodes that need to be plugged into the lines of communication formula:
$$lines\ of\ communication = \frac{n(n-1)}{2} = \frac{31(30)}{2} = \frac{930}{2} = 465$$

10. Answer: C. Although activities A and B do not depend on one another, they both have a relationship to C. A must be done before C can begin and C only begins when B is complete. Since the problem states they take the same amount of time, A and B ought to begin at the same moment, so their relationship is start-to-start. Since A is tied to quality control of B, A is the successor activity in this scenario. SF relationships only exist when the successor activity must be moved before its predecessor due to a delay or resource constraint.

Chapter 3: Executing

Section 3.1: Teamwork and communication during the project

All the planning processes outlined in the previous chapter pay off as the project is executed. During the execution phase, the majority of the budget is spent and most problems are encountered. Strategies for dealing with many of these problems will be addressed in the next chapter, Monitoring and Controlling. Nonetheless, as the plans for a project should have in-built flexibility and protocols for problems, planning strategies from the previous chapter will also come up during the execution phase.

Consistency on good and bad days is the most important factor for a smooth execution phase. Team-members should know the project manager's preferred ways of doing things and know what to expect. The project manager should be dependable, never amplifying problems. **Management by exception** is the term that best captures this attitude by instituting standard processes to respond to each type of issue. Managing by exception focuses on small breaks with project plans to prevent larger issues from occurring. Managing by exception comes up most when communicating to a project team while they are delivering on the project's goals. There are two groups within the project with whom a project manager should prioritize consistent communication and best practices: project drivers and key stakeholders.

During the planning and execution phases, project drivers' voices should be heard. If the project is using a RACI (responsible, accountable, consulted, and informed—as covered in chapter two), drivers are the people most listed as "responsible" or "accountable." During weekly meetings or emails with large portions of the project team, drivers' needs and opinions about project progress should be heard. This helps further embed a one-team, holistic view of the project. If drivers and more subordinate team-members know each other better, communication between those two groups will increase. Facilitating open communication also ensures that you will have an accurate picture of what is going on and not have your team hide errors from you.

Especially during a project in which the execution approaches a year's time or more, stakeholders may lose contact with the project. That would be a setback, as stakeholders often shape how the buyer or client feels about the project. Stakeholders should be invited to offer feedback at every stage of a project and their expectations should be common knowledge. Engaging stakeholders and surveying their opinions is often a straightforward way to avoid dissatisfaction at a project's completion. A project manager that understands stakeholders' **WIIFM** ("what's in it for me") perspective is most likely to satisfy widely held, but perhaps not explicit, wants in the project. Often, stakeholders will appreciate being invited to voice their opinion and will be less likely to criticize a product they feel captures their views.

Even if project drivers or stakeholders are only interested in communicating about a small portion of the project, communication should be put in writing as often as is possible. Formalized communication that documents your efforts at gathering opinions, as well as agreed-to plans, is more likely to get a response than a perfunctory phone call or, worse, a text message. As the project's end date approaches, a project manager will feel more confident that everything has been aired out if it is in writing. Communication should be specific and direct, with exact dates and clear writing. The sender is responsible that the recipient understands the message. Active voice and precise language are good starting points in written communication. If something is being requested, write it straightforwardly and follow-up on messages if a reply is not forthcoming. Even if talking to a superior, a good project manager should take a long-term view and respectfully pester them for an answer to make project success more likely. Beyond best practices, getting written approvals about vendor orders and technical specifications are often a requirement for many project types.

Communications dealing with large project activities should be documented in an **issue log**. There should be categories for senders, receivers, and a message summary. Stakeholder outreach should be marked. There should be an orderly system in keeping tabs on messages and getting replies. The issue log is used for far more than communication reporting, and should include categories on severe risks and other holdups. Nonetheless, tracking communication is as important as any other log-entry. Regarding communication, the log should also separate out the following three types of communication: lateral communication, upward communication, and downward communication.

Lateral communication (or cross-organizational) is conducted through peers or near-peers, such as a project manager, contractor, and functional manager in another department. Those communicating laterally are heavily invested in the project. **Upward communication** consists of a project manager reporting on project progress to executives or influential stakeholders. These should occur throughout the project in a uniform manner and increase when problems arise. **Downward communication** occurs within the project team and concerns day-to-day tasks. New assignments, schedule updates, and reviewing others' work are the most common topics in downward communication. The **controlling communications plan** is the subsidiary plan that structures how a project manager will conduct themselves through each type of communication: lateral, upward, and downward.

The choices a project manager makes when communicating with everyone involved in the project is one of the main sources of authority a project manager has. Authority is the ability to make binding decisions within the project. Deciding to email someone at a particular time or leaving someone off of a message can have large ramifications. Responsibility and accountability are different. Responsibility entails a commitment to achieve a goal, while accountability has to do with handling consequences. A project manager can delegate authority for certain tasks, but not for communication. You should be at the center of every message; copied on every project team email. No matter what happens, a project manager should never be caught unawares after something occurs. If so, they have failed in their communication strategy. The project manager should be the first person to know when something goes wrong, beyond the people experiencing it first-hand. Understanding this responsibility and accountability is a large part of the ethical standards involved in being a certified project manager.

To ensure that your team reports and communicates as desired, many project managers are well-versed in motivational theories. Abraham Maslow's **hierarchy of needs** is foundational and stipulates that people need to feel they are actualizing their potential to be truly invested in an endeavor. Self-realization is the highest need, and more important than needs based on physiology, safety, belonging, and esteem. Frederick Herzberg's lesser-known and oddly-named **motivation-hygiene theory** looks at two motivational theories. Motivational factors deal with the satisfaction garnered from doing the work and mastering the many skills required in doing that work. Hygiene factors deal with the environment, and are broadly understood to include things that prevent dissatisfaction, such as pay, social ties with team-members, and a pleasant work setting. Herzberg believed that if people expect to be rewarded after completing an assignment they would do it better and motivation is external. David McClelland's **needs theory** hones in on achievement, power, and affiliation as three things team-members need to believe they possess to dedicate themselves to a project. McClelland's model posits motivation is internal.

Douglas McGregor developed a model for two types of managerial practices in motivation that map onto the motivational theories above. **Theory X managers** are hands-on. They believe people avoid working hard and take short-cuts. Theory X managers stress discipline and rules and subscribe to the motivation-hygiene theory, doubting their workers want to succeed because they like the work. As you would imagine, **Theory Y managers** are the opposite, and hands-off. They believe people work up to expectations and outperform when held accountable. Theory Y managers subscribe to the needs theory and do not micromanage.

A theory Y style is similar **to laissez-faire leadership**. Laissez-faire (meaning "let them do as they please") project managers delegate their authority and do not get involved day-to-day. William Ouchi added a nuance with **Theory Z managers,** which is more about the setting than external or internal motivation. Ouchi saw motivation stemming from then environment and job security. If team-members feel confident that their work is stable and that they are free to take the occasional risk, their morale would be higher and performance improved.

These three types are idealized and not individually attainable within many projects. The best option is a mix of Theories X and Y, and to keep Z in mind if it's applicable to your workplace. When conducting team-building activities to strengthen your team, it is wise to keep the hands-off and hands-on dichotomy in mind. If you favor one method too much and are not getting the results you want, you should try changing it up during the next project and introduce new ground rules. However, for the sake of consistency, there should not be drastic management-style changes during one project. Having a working theory on your team's work habits will assist your interpersonal skills, a highly important trait during project work.

Adding new team-members could change the dynamics of your team and lead to new results to tried-and-true methods. Whether it is the acquisitions department in a large organization or a smaller-scale hiring operation, there are best practices to ensure you hire someone that matches the project's ethos who also amplifies existing positive trends. **Multi-criteria decision analysis** helps determine the likelihood of a new team-member jelling with your existing staff. The criteria vary by the project, but experience, education, work-setting preference, and the weekly hours target are typically considered. The chosen criteria are used to rank candidates from best to worst. Competency, often used as one of the multiple criteria to analyze recruits, refers to a specialized skill attained from experience or education.

Resource calendars are developed during the planning stage and guide the execution phase's development. Resource calendars display team-members' assignments and when they are scheduled to work on project activities, as well as when non-human resources will be available. For large projects, the availability of certain competencies during project stages will be marked as well. Corrective actions reallocate resources and edit calendars in order to return the project to the planned timeline as soon as possible.

The location of the team and proximity of members to one another will determine a lot in how a project manager handles the team and its workflow. Teams in a **tight matrix** are those that are co-located and work with each other daily in the same physical space. A tight matrix can take the form of a weekly meeting where everyone is physically present, or daily interactions of some sort in a shared space. Tight matrix teams must see each other continually in a sustained fashion. **Remote teams** are separated and communicate virtually.

Team-members in this case are still committed to the project, and many work on a daily basis. With the increasing pace of communication capabilities, this arrangement is common and allows for a more diverse set of specialists to collaborate. Often, tight matrix co-locating teams require a different set of supervision compared to remote teams. Remote teams often lose sight of the total project goal and it is incumbent upon the project manager to remind everyone how they fit together.

Whatever the properties and quirks of your team are, **norming** is an essential stage in ensuring continued project progress. Norming develops guidelines to regulate team-members' behavior and can be formalized in a handbook, or informal and mentioned in team meetings. Norms are best instituted at the start of a project and followed throughout the project. Things such as when to send a second, reminding email, to presentation protocol and how to ask questions, all count.

Conflicts are most likely to erupt when norms fall into disuse or are transgressed for whatever reason. When a team member expects a certain thing from another member and that person fails, usually the project manager must step-in to resolve things before there is conflict escalation and communication shuts down. The project manager must step in and engage in **conflict management**. There are five common conflict management methods, ordered below from those that take the least to most time:

> - **Withdraw** implies a conflict is ignored rather than addressed. This could occur when one party refuses to discuss the conflict and merely adjusts, a party leaves the project, or the conflict was a minor issue that does not warrant attention.
> - **Smoothing** tries to create breathing space in the short-term to see if the issue blows over. It minimizes problems and tries to reinforce ground-rules in communication that may have lapsed.
> - **Direct force** doles out a decision to address underlying causes for team friction and demands compliance. While not the best strategy, direct force is often resorted to after repeated offenses and when deadlines are approaching. Usually, the underlying cause is not addressed, and the conflict pops up again.
> - **Reconciliation** attempts to have aggrieved parties compromise on something, share the blame, and agree on a solution. For example, one party can admit they had unrealistic timelines the other did not consent to, and the other party can admit they did not respond up to their own professional standards as communication slowed. This often requires the project manager's intensive understanding of the issues at play and an investment in time to resolve the problem.
> - **Problem solving** requires the project manager to introduce a new set of guidelines to address underlying issues permanently. Voices beyond those implicated in the conflict are involved and a solution is

constructed from those multiple opinions. The existing conflict is addressed in hopes similar problems in the future will be prevented from a solution derived from consensus.

These conflict resolution techniques can be understood along a grid, in which the conflict is either not escalated or escalated, to the point in which more and more people are involved. Usually, a problem is escalated if it involves individuals that are accountable within multiple deliverables. The more norming, especially in communication protocols, and team-building, i.e. motivational thinking, that goes on earlier in the project, the more successful conflict management will be later on.

The **Thomas-Kilmann Conflict Mode Instrument (TKI)** offers another way to achieve team cohesiveness. The TKI uses a questionnaire to determine how team-members act in tense situations. With this data in hand, project managers can adapt conflict management techniques to their team to resolve issues. If a project manager's conflict resolution efforts are not successful, then an **escalation of issues** is necessary whereby higher levels of management are involved. Escalation of issues occurs any time during the execution of a project in which a project manager cannot get things back on track.

Section 3.2: Working through the schedule

Keeping track of project progress relative to the schedule, and making updates as work is completed, will take up a lot of time. The project's baseline represents the original plan for the project—including the cost baseline, scope baseline, and schedule baseline. The project management plan's many subsidiaries inform these baselines. As the project plan is executed, there should be a daily check on how the project is or is not aligning with the plan.

For all the planning that goes into the schedule baseline, the **level of effort (LOE)** estimate is likely to cause the most headaches. LOE is used to determine how long an activity will take and depends on the skillset of the team-member. It is assumed experienced individuals would take less time than a novice. However, with sicknesses, departures, and unexpectedly good and bad performances, there will be a wild inaccuracy in LOE in every project. If a minor activity is supposed to take one team-member one week, but ends up taking three people's time during that one week, the LOE turns out to be three weeks rather than one.

Since the project manager will be in charge of monitoring the baseline, LOE changes, and the bulk of the project's communications, they should avoid assigning themselves many tasks. Project managers are often a reserve resource to step in when LOE estimates are inaccurate. Especially in smaller firms, the project manager often does double-duty and manages the project while also being responsible for the occasional project task. If this is necessary, avoid assigning activities to the project manager's workload that are on the project's critical path. If the project manager is responsible for critical path activities and stepping up when more effort is required, delays are far more likely to occur. A project manager should assign themselves as few activities as possible.

As the project manager oversees schedule, the following six properties guide decisions about moving things around:

> **Total float** refers to the amount of time an activity can be postponed and begun later without pushing back the overall project end date.
> **Free float** refers to the amounts of time an activity can be postponed without delaying the start of another separate activity.
> A **lead** modifies a relationship between project activities and accelerates work on a dependent successor task when the preceding task is bogged down.
> A **lag** extends the time between two dependent tasks. A lag is necessary when two preceding tasks feed into one successor task and one of these preceding tasks is completed much earlier than the other.
> **Crashing** shortens the entire project's timeline by consolidating activities and adding more resources. Activities that take the most time are focused on, and their end-dates are brought up. Crashing adds float relative to the original timeline.
> **Fast-tracking** requires that two tasks originally scheduled to be done consecutively are instead performed at the same time.

The importance of managing the starts of tasks is enlarged when the budget is tied to the project's progress through deliverables. It is often the case that funds are only released when prior project activities are completed. A **time phased budget** entails such co-dependency and is a common condition in a project contract. Money is only released when a number of tasks are completed. If there are multiple delays involving many deliverables, the project could find itself in the red. The **undistributed budget** is the total sum of funds that have yet to be allocated on a task-by-task basis. If the schedule has to be altered, the expected undistributed budget breakdown has to be reworked.

Another technique to assist a project manager in their corrective scheduling is using learning curves to estimate. Learning curves represent the time it takes team-members to learn a task. The guiding principle is that the higher the frequency of the task being performed, the less time and money is required on a per-hour basis. When labor is a primary resource in a project, such as in a manufacturing one, learning curves are especially pertinent, as economies of scale and other efficiencies are realized. With technology and deep learning (machines' algorithmic adaptions) becoming a large part of the workplace, learning theories and their benefits are likely to proliferate.

The exact benefits of using learning curves are sometimes difficult to ascertain beforehand. Relying too much on a theory presents many dangers. Every project should be based on empirical observations. An empirical project manager relies on data obtained by observations and experimentation in a controlled environment that approximates your specific project environment as much as possible. The opposite of this would be **confirmation bias**, which is the tendency to interpret results to support a pre-existing viewpoint, rather than being open to being wrong.

When estimates and plans do go awry, **corrective actions** must be taken. The actions will realign project progress with an updated project management plan. It's worth keeping in mind that corrective actions can be needed for positive and negative reasons. There are situations in which hidden efficiencies are discovered and the project can be fast-tracked. Preventive actions and contingency plans are sometimes a necessary complement to corrective actions. A preventive approach tries to contain spill-over into other project activities, reducing the likelihood of similar errors from recurring and limiting the resulting damage if the same sorts of errors do reoccur.

Changes often require trade-offs in which one activity loses resources and time to another. **Trade-off analysis** aims to evaluate the effects of moving resources away from one activity to another in order to optimize the upside of a trade-off. Most trade-offs are competing ones: changing the resource distribution between competing activities invariably helps one and hurts the other. It's a zero-sum situation where there is a finite amount of resources. Complementary trade-offs, meanwhile, do not compete for resources, but rather increase the total amount of resources in the project by finding efficiencies. Competing trade-offs divide a pie into smaller slices, while complementary ones grow the pie.

Changes can come from stakeholders asking for an enhancement, clients limiting the budget, new efficiencies, and a variety of other sources. **Total quality management (TQM)** is a technique developed by Armand Feigenbaum that tries to introduce all types of project changes smoothly and with as little disruption as possible. TQM brings in all team-members and encourages preparation for and understanding of changes that are about to be introduced. Being aware of errors in change-implementation is emphasized. The TQM method should be employed to monitor the project, especially when trade-offs occur.

Section 3.3: Guidelines for contracting

This section will deal with issues that often are first encountered in the initiation phase, but very much shape the execution of the project. It will be especially useful to project managers likely to work on projects that are regulated by contracts between two or more separate organizations. Procurement and contracts rules are of such centrality to the project management profession that project managers are expected to be well-versed in the details outlined in this chapter.

Every project should have a procurement process to identify the purchases the project needs as it progresses and how team-members can request additional purchases. **Make or buy analysis** is a critical step in the procurement process. Many large firms have the resources to "make" many goods or services in-house, while smaller firms have to "buy" more outside help. But the decision to make or buy is a decision about whether to initiate the procurement process. Procurement plans should define which resources will be received internally within the organization and which resources will require an outside partner.

The **procurement statement of work (P. SOW)** contains all of this information and is originally completed in the planning phase. The final P. SOW is a negotiation between buyers and sellers that have partnered to provide a good or service to a project. The statement goes through multiple drafts and revisions until both sides are satisfied. If one of the parties is new to the type of project under consideration, the most experienced side should take the lead and prepare the first draft. Since some of the resources and associated contractors will not be part of the project until the execution phase, the P. SOW may be revised substantially as work gets under way.

In larger organizations, the procurement department should estimate the costs of every procurement. In smaller organizations, the project manager should be committed to getting the best value and attempt accurate cost estimates. **Benefit-cost ratios (BCR)** indicate the total per-dollar value of a proposal and can be used throughout the project's life anytime an item of expenditure is being considered to compare options. BCR is often used when considering outside contractors. The ratio is found by dividing the proposal's cash benefits by its costs. Net-present value, covered in chapter one, helps calculate the cost benefits. When comparing BCRs and trying to mix-and-match good qualities from multiple offers, a project manager's negotiating skills are tested.

Procurements can be attained by issuing requests for proposal (RFP), requests for information (RFI), invitations to bid (IFB), and requests for quotation (RFQ) to suppliers. Once these documents are issued and publicized on the appropriate website for the line of work, vendors or suppliers respond and it is up to the project-team to consider the offers. Entire projects begin this way, but so do small jobs within a project that require specialized skills or equipment. **Preassignment** skips, or at least expedites, the process of requesting and reviewing outside firms in the procurement process. Often, a buyer will have a partner organization in mind and offer them the contract first. When this occurs, specific individuals are often requested with whom the buyer had a positive prior experience. On the other end of possibilities, contractor conferences allow vendors of services to meet buyers for the first time. For many industries, there are many prospective sellers for each buyer and contracts usually have many bidders.

Even with preassignments, a project manager will be on the frontlines of negotiations when agreeing on a contract. Negotiating is mainly honed with experience. One project itself usually contains enough teachable moments to turn a project manager into a grizzled negotiator. Although often beginning in the initiation phase, the execution phase and the accompanying changes witness an intensification of negotiations on the fly. The type of contract agreed to dictates much of the relationship between contractors and the contracting organization. Availability and work hours are often one of the stickier things to iron out. The specific dynamics, however, flow from the following contract types and their subtypes. **Contract types** are all legally-binding agreements that determine the level of work and risk obligations for the seller and buyer during a specified period of time.

> ➢ **Fixed price contracts** are mostly employed for detailed projects with few risks or where there is a high level of trust between the concerned parties. There are three types of fixed price contracts: *Firm Fixed-Price* flatly defines the cost of a deliverable and expect the seller will incur any extra cost; *Fixed-Price Incentive-Fee* includes a bonus amount if a certain goal is met regarding timeline or quality; and a *Fixed-Price with Economic Price Adjustment* ties the costs of the price to a financial index, because the specific costs are determined by market fluctuations.
> ➢ **Cost reimbursable contracts** are meant for projects with high levels of risk and with many obstacles. There are defined and funded costs, usually towards the start of the project, but there is a shared understanding that the costs in the latter phase of the project are uncertain. All cost reimbursable contracts see the seller receive periodic credits from the buyer for all charges. There are four types of cost reimbursable contracts: *Cost Plus Fixed Fee* contract allows the seller to charge the buyer every single cost of a project and the seller receives a fixed fee (the seller's profit) at project completion; in a *Cost Plus Incentive Fee Reimbursable* contract, the seller's profit is tied to an incentive dependent on a timeline or quality; in a *Cost Plus Percentage Of Cost Variant* contract, the seller's profit is an agreed-upon percent of the cost of the project; and the *Cost Plus Award Fee* variant allows the buyer to unilaterally decide what to pay the seller (the seller's profit) when the project is complete, after all costs are covered.

No matter the type of contract binding a seller and buyer, there are commonalities to all contracts in nearly every legal environment. The **force majeure exemption** allows a contract's obligations to be nullified when outside conditions prevent the contract's work from being undertaken (an "act of god"). The seller's costs in the scenario are reimbursed, but the project is never completed. The **privy mutual relationship** specifies that the requirements of a contract only apply to the parties specified in a signed contract, not to outside organizations.

Whenever there are multiple vendors involved in a project, both the buyer and seller are responsible for coordinating work between each party. Every part should be motivated to stay on track for a timely completion of a project. A contract usually involves a procurement agreement with specific reporting requirements. Reporting on project progress is usually the sole responsibility of the seller. The buyer is responsible for providing feedback and raising concerns. A project manager is at the center of this communication and must force the parties to resolve issues, especially if a subcontractor or the buyer is slow in answering communications.

If a contract must be modified, the project manager must be cognizant of the type of contract defining the project and the legal recourse for modifications. The strategies for schedule change outlined in chapter two apply to contract modifications as well. Contract modifications are known as amendments. If the contract cannot be amended and there is an impasse, the contract can end in one of three ways. **Ending a contract for cause** is when one party breaks the contract's rules and is unable to hold the terms. **Ending a contract for convenience** is when one party to the contract withdraws from the contract by choice. **Ending a contract by default** is due to time expiring and the product or service to be rendered is no longer needed by the buyer.

Section 3.4: Risk management

Although dealing with risks is a process that coexists with the entire life of the project, strategies for handling risk often matter much more during the execution phase of a project. The strategies are set in the planning stage, though they are carried out and modified during project execution.

A **negative risk** is a threat that will possibly impede project progress, such as losing out on a key resource during a deadline. If a negative risk occurs, it becomes a live issue that must be addressed and is no longer a risk. A **positive risk** is an opportunity to possibly achieve an objective that will help the firm, like completing a project with less resources. If a positive risk is determined worthwhile and the opportunity works out, it is no longer a risk, but a project asset. **An exploit strategy** is executed when there is a positive effect to a risk occurring, such as monetary savings, and acting in such a way to realize that positive outcome. **A secondary risk** arises in response to decisions made about a prior risk, such trying to exploit it.

When trying to determine the nature of a risk, it is important to analyze the risk from the following perspectives within your organization:

> ➢ **Risk appetite** details the level of unpredictability project leaders can accept based on resources. Project decisions and gambles are

determined by how great the risk appetite is and the corresponding budgetary flexibility.

- ➤ **Risk tolerance** details the level of unpredictability project leaders can accept based on risk-reward calculations. Tolerance will be higher if project leaders reason that the reward possibilities outweigh the potential costs.
- ➤ **Risk thresholds** define the range within which project leaders are willing to take a chance. Certain risks could result in unacceptable costs. Some risks pose costs potentially so great, they are beyond risk thresholds and therefore will never be attempted.
- ➤ **Insurable risk** is negative eventualities in which an insurance policy can cover part or all of the costs (weather, equipment, currency fluctuations, etc.).

Often, firms will seek to minimize risk and exposure. **Risk transfer** occurs when the risk has not occurred and the potential consequences are taken up by a third-party. Purchasing insurance for various possibilities is a form of risk transfer. Another is purchasing hours in advance from a subcontractor that may or may not be used based on risk outcomes. If a firm becomes more risk-averse as a project goes on, the budget will have to be altered in order to account for possible risk transfer expenditure. For organizations with copious amounts of capital and investments, risk transfer will take up a large portion of a project's plan. For example, project managers involved with the airline industry will experience that industry's financial maneuvering to hedge on the price of petroleum for example, a form of risk transfer.

A **share strategy** is a form of risk transfer that deals with comparative advantage and turning risks into opportunities. It again involves outsourcing risks to a third-party. Rather than the motivation being insuring and preventing future loss, the partnership in a share strategy is to increase overall revenue. If you are a manufacturing firm that wants to develop a product for a new international market but have little experience with the export laws to the country in question, you may share the related risks of the project with a firm that is familiar with working in that country. In this scenario, the firms are complementary. Risk transfers have to do with limiting risks in general, while share strategies focus on the type of risk and the limits of principle of the firm executing the project.

Handling risks and the fallout often entail involving far more than just those involved in a project. A negative risk occurring usually leads to a need for more funds and therefore brings in many individuals superior to the project manager. Being proactive and preventing the unexpected from occurring, and then containing the fallout, should be a top priority in any project. Spelling out roles and responsibilities and knowing who to contact for risk categories is a crucial step to control risks. Risk teams do not have much overlap with project teams.

The best thing a project manager can do to contain risk fallout is to construct and maintain a **risk register**. The risk register is a list of risks ordered by priority and including each risk's probability, likely cost, and timeline. Risk analysis must be unbiased and holistic, playing out for many steps the costs of a breakdown in every facet of the project. **Workarounds** are responses to unexpected events that are not planned but required to solve a pressing issue or negative risk event. Often, it is up to a project manager to decide on the specific workaround. By definition, workarounds are not included in the risk register since they are unforeseen.

Risk tracking encompasses the methods a project manager will employ to record risk activities. Developments are logged before the threat materializes, and tracking will include actions to minimize costs. If your firm is a startup, you may have to come up with your own risk tracking system. There are multiple online programs to assist risk tracking.

Triggers or risk symptoms are indications that a risk will materialize into a threat. The triggers a project manager will look for can be both small and big. A small trigger can be a subject-matter expert taking an inordinate amount of time to reply to an email. These delays could be an early indicator of a delay in the associated deliverable. A big trigger would be early project activities costing far more than expected. This budgetary indicator could foreshadow a major understaffing in the project team and trigger an appropriate action to prevent the risk from occurring.

In terms of project teams and risk teams, there should be communication at the early stages of a trigger. Team-members should discuss the nature of the issue and propose solutions. The risk register and tracking system should record all of these conversations. It is important to come to a consensus as quickly as possible and agree on a solution to address at least some of the more basic problems associated with the risk. As the risk develops, it may be necessary to redo some of the risk register's categories to account for the changing nature of the risks that threaten the project. More granular categories are often required as a project matures. As triggers accumulate, it will likely be necessary to recategorize risks and change the probabilities of many associated risks.

All risks can be sorted by the following types, which share similar characteristics no matter the project specifics:

> **Project management risks** stem from poor scheduling, budgeting, and inconsistent methodology. These risks arise when a project plan is found to be extremely inaccurate during the execution phase.
> **Organizational risks** stem from in-fighting and competition for resources within an organization that often is juggling multiple projects. The larger organization rejects the project's scope and budget once the project begins and needs resources. Organizational issues can also stem from a

lack of logistical knowhow when it comes to fulfilling project requirements.

➢ **Performance risks** stem from an organization depending on untested methods, people, corporate partners, or technology to fulfill project requirements. Performance risks can also be labeled as technical or quality risks.

➢ **External risks** have nothing to do with the project. Changes in the larger environment surrounding the project, like stock price fluctuations, new laws, weather, power outages, internet connectivity, and the project being terminated for a reason not having to do with the project team.

➢ **Product risk** occurs when a new product, including software and hardware products, fails to live up to expectations. This failure may be due to scheduling issues with unexpected technical slowdowns, or resource issues due to either compatibility or capability shortcomings in equipment and facilities.

Regardless of how well risks are anticipated and analyzed, some organizations are more risk-averse than others. **Maximin Criterion** (or Wald Criterion) emphasizes the losses possible in risks—it is the "half-empty" view. **Maximax Criterion** (or Hurwicz Criterion), however, emphasizes the possibilities inherent in a risk—it is the "half-full" view. Having a solid sense of an organization's attitude about risk will limit a project's risks and performance risks will be easier to head off. In an extreme maximax organization, positive risk events are exploited to ensure they happen and the associated changes take place. Exploiting risk events in such a scenario is a large gamble.

Expected value is how both maximin and maximax organizations decide how to approach and try to figure out the positives or negatives associated with a risk. Expected value is the numerical product of probability a risk materializes and the impact of the risk. The equation is:

$$expected\ value = probability\ x\ impact$$

Probability is assessed between 0 (no probability) and 1 (100% certainty). **Impact** is the amount of harm or help a risk carries. Impact is expressed on a scale from 0 (very little effect on the project) to 1 (critical effect on the project). Impact's exact meaning varies and is decided by the organization. For better calculation, it is important for there to be gradual increments in impact values— for there to be an impact at .2, .4, .6. and so on corresponding to distinct results. Expected values closer to zero are low-level risks, while expected values closer to one are high-level risks. Ranking risks' expected value from those close to zero to those close to one is known as **ordinal scale**. The following graph illustrates the fundamental ideas and uses of expected value calculations:

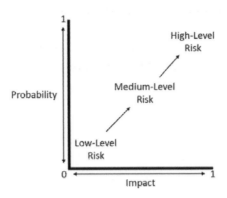

For example, if the probability of a risk even occurring is .2 (20%) and its impact is assessed at .25 (25%), graphing those two values will yield a "low-level risk." A project manager should isolate risks that appear to have high probabilities and impacts and get it on the radar. With the project manager, the appropriate people can make decisions on either embracing the risk, or taking action to prevent the likelihood of the risk occurring, depending on the risk profile of the organization. Separating low-level and high-level risks is a key to effective risk management.

Impact can also be measured in hard currency. If impact is measured in dollars, for example, a firm can find the **expected monetary value (EMV)** of an opportunity or risk. The EMV is an amount that is either a profit, a positive number, or a loss, a negative number. EMV is often used to assess projects during the initiation phase. It can also be used to evaluate decisions with a project. The EMV is found by subtracting the probable expected loss from the probable expected profit. The equation is below:

$$EMV = (profit\ probability)(expected\ profit\ amount)$$
$$- (loss\ probability)(expected\ loss)$$

In addition to understanding a risk's nature, it is imperative to have plans in place to deal with all types of risks. While the planning chapter above outlines many strategies to head off risk, the techniques below are more strategies that fall more in the domain of risk-management.

Most techniques to handle risk are either a form of contingency or mitigation planning. **Contingency planning** recommends changes to make in response to a risk that occurs. Contingency plans take the form of: "if this, do this." **Mitigation planning** offers strategies to implement in order to make a risk less likely from occurring or to contain the risk's effects if things start to worry the project manager. In terms of the expected value equation above, contingency planning focuses on shaping impact, while mitigation planning focuses on shaping probability. Both types of strategies are altered and strengthened as a project is being executed.

Reserves are resources or assets that are meant only to be used in the event of a risk occurring. Reserves are sometimes built into a project's budget to assist in risk-management. But reserves do not have to take a monetary form. In a large organization, for example, some team-members may be held on reserve and only placed on a specific project if that project runs into a specific sort of problem requiring a skillset not often needed for the project.

So much of handling risks has to do with analysis. Risks are fundamentally a failure of understanding the nature of a project. The following are tested methods to better analyze a project's risks that can occur during any time in a project:

- ➢ **Brainstorming** brings together project team regulars and irregulars, including executives and third-party contractors, to map out the likeliest risks and obstacles a project will face. Project managers should try to bring the team together early on for this sort of activity since it may not be until a problem is encountered that this sort of collaboration is needed.
- ➢ **Interviewing outside specialists** can help hedge against future risks. Especially in an area new to an organization, it is wise to get many opinions about similar projects. A straightforward way to get such advice is during the solicitation of contractors. They probably played a part in similar projects in the past, which is why they are again under consideration to handle certain tasks, and can provide good information on bumps during project execution.
- ➢ **Checklists** can help any project manager avoid careless mistakes that exacerbate risks, as well as ensuring consistency over a variety of scenarios. Checklists are built and modified based on prior work history and consultation with specialists.
- ➢ **Root cause analysis** tries to look past symptoms of repeated failure and find perhaps unnoticed issues. Team chemistry, communication protocol, or organizational weaknesses are all fair game in this approach.
- ➢ **Delphi technique** uses outside experts independently to get a fair assessment of issues in the project or the project plan. Specialists are contacted and asked to develop an analysis of a project's strengths and weaknesses when it comes to likely risks. Experts do not know who the other analysts are or if there are other analysts at all. The results from each analyst are then compared and similarities in each report are acted upon.

When risk-analysis is faulty, specialized organizations can help by conducting audits. **Risk audits** are continuous checks on the project's progress from start to finish to identify upcoming obstacles or ongoing underperformance. Auditors should have access to the project manager to keep tabs on the project.

Section 3.5: Practice exam questions on executing

1. An issue log should have which of the following characteristics?

   a. Categories for communication with stakeholders and team-members
   b. Categories for lateral, upward, and downward communication
   c. The text of each message sent to a stakeholder
   d. All of the above

2. Which of the following scenarios correctly describes the differences between theory x, y, and z managers?

   a. While theory x and y managers focus on workers, theory z managers focus on goals
   b. Theory x managers have a negative view of workers, theory y managers have a mixed view of workers, and theory z managers have a positive view of workers
   c. Theory z managers have a negative view of workers, theory x managers have a mixed view of workers, and theory y managers have a positive view of workers
   d. While theory x and y managers focus on workers, theory z managers focus on setting

3. How is successful norming related to a project team in a tight matrix formation?

   a. Norming develops guidelines to regulate team-members' behavior. A handbook could be employed to institute the rules even more formally. A tight matrix team works in proximity to one another and norming emerges from regulated, daily interactions that the project manager can oversee.
   b. Norming develops guidelines to regulate team-members' behavior. The relevant rules are more informal and change frequently. A tight matrix team works in proximity to one another and norming emerges from regulated, daily interactions that the project manager can oversee.
   c. Norming develops guidelines to regulate team-members' behavior. The relevant rules are more informal and dictated by work culture. A tight matrix team works remotely from each other and norming must be stressed by the project manager from afar.
   d. Norming develops guidelines to regulate team-members' behavior. A handbook could be employed to institute the rules even more formally. A tight matrix team works remotely from each other and norming must be stressed by the project manager from afar.

4. A project manager at an artificial intelligence firm realized the project veered far from its baseline. The share strategy the firm used with a marketing firm led to unexpected delays. To get back on track, the project manager seeks and gets approval to consolidate activities and add more resources to the project. Which of the following describes the strategy the project manager employed?

   a. Total float
   b. Free float
   c. Crashing
   d. Fast-tracking

5. Which of the following is not a fixed-price contract?

    a. Fixed-Price Incentive-Fee
    b. Fixed-Price with Economic Price Adjustment
    c. Cost Plus Fixed Fee
    d. Firm Fixed-Price

6. What is the expected monetary value of a risk transfer scheme that has a 30% chance of resulting in $12,000 more profit but a 50% chance of resulting in a $5,000 loss?

    a. $1,100 profit
    b. $7,000 loss
    c. $2,000 profit
    d. $1,100 loss

7. Which of the following is true of project risk management?

    a. Stakeholders should be involved early on and know about potential risks
    b. Risk triggers should be identified after a risk event occurs
    c. Risks should be rated by priority in the closing process group
    d. All of the above

8. When should a project manager pursue a workaround?

    a. After getting permission from upper management to change project plans
    b. After updating the risk register, change plans accordingly
    c. After letting the entire project team know about the new plan
    d. After an unexpected risk emerges for which no plan exists

9. Requests for proposal (RFP), requests for information (RFI), invitations to bid (IFB), and requests for quotation (RFQ) all have what in common?

    a. They all lead to fixed-price contracts
    b. They all lead to cost reimbursable contracts
    c. They are part of the procurement process
    d. They are all part of the preassignment process

10. What is the relationship between an exploit strategy and a secondary risk?

    a. An exploit strategy is executed when there is a positive effect to a risk occurring, such as monetary savings, and acting in such a way to realize that positive outcome. A secondary risk arises in response to decisions made about a prior risk, such as an exploit strategy.
    b. An exploit strategy is executed when there is an opportunity to transfer the risk outside of the organization. A secondary risk arises in response to decisions made about a prior risk such as an exploit strategy.
    c. An exploit strategy is executed when there is a positive effect to a risk occurring, such as monetary savings, and acting in such a way to realize that positive outcome. A secondary risk is the level of risk that is accepted in the project.
    d. An exploit strategy is executed when there is an opportunity to transfer the risk outside of the organization. A secondary risk is the level of risk that is accepted in the project.

Section 3.6: Practice exam questions on executing – answers

1. Answer: B. Issue logs document communication pertaining to large project activities. There should therefore be a category for communication with stakeholders. A category for team-members is too broad. Communication with team-members is either upward or downward, depending on the authority of those in question. The text of each message is too exhaustive. Lateral communication is cross-organization. There should be a summary of messages that are documented. Just a sentence usually is fine: "reminder that a deliverable is due in two weeks."

2. Answer: D. Theory x managers are pessimistic about worker motivations and stern and theory y managers are more optimistic and believe workers can be self-starters. Both focus on workers, while theory z focuses on the larger work setting.

3. Answer: A. Norming does indeed develop guidelines that are better formalized than not. A team in a tight matrix formation works in the same space and interacts in-person daily. Answers C and D are referring to a remote team.

4. Answer: C. Much of the information in this problem is a distraction. Any time a "project manager seeks and gets approval to consolidate activities and add more resources to the project," they are crashing. Total float refers to the amount of time an activity can be postponed and begun later without pushing back the overall project end date. Free float refers to the amount of time an activity can be postponed without delaying the start of another separate activity. Fast-tracking requires that two tasks originally scheduled to be done in order are instead performed at the same time.

5. Answer: C. A Cost Plus Fixed Fee is a cost reimbursable contract. Cost Plus Fixed Fee allows the seller to charge the buyer every single cost of a project and the seller receives a fixed fee (the seller's profit) at project completion. All of the others are fixed-price contracts. Fixed-Price Incentive-Fee includes a bonus amount if a certain goal is met regarding timeline or quality. Fixed-Price with Economic Price Adjustment ties the costs of the price to a financial index because the specific costs are determined by market fluctuations. Firm Fixed-Price flatly defines the cost of a deliverable and expects the seller will incur any extra cost.

6. Answer: A. Plug in the correct amounts into the EMV equation. Since the final number is positive, there is an expected profit.

$$EMV = (profit\ probability)(expected\ profit\ amount)$$
$$- (loss\ probability)(expected\ loss)$$
$$EMV = (.3)(12{,}000) - (.5)(5{,}000)$$
$$EMV = 3{,}600 - 2{,}500 = 1{,}100$$

7. Answer: A. Stakeholders can provide insights to help mitigate risks and should be involved early in risk identification. Risk triggers should be identified before risk events occur. Risks should be rated by priority in the risk register.

8. Answer: D. Workarounds are responses to unexpected events that are not planned but required to solve a pressing issue or negative risk event. Often, it is up to a project manager to decide on the specific workaround. By definition, workarounds are not included in the risk register.

9. Answer: C. These requests are the typical ways an organization finds a contractor from which to procure desired services. The resulting contracts can be fixed or reimbursable. Preassignment is the opposite of requests. Preassignment occurs when a firm has a certain contractor in mind and skips the requesting and call stage.

10. Answer: A. Only the first option has the correct definitions. Risk transfer occurs when a firm outsources risk to another organization. Residual risk is the level of risk that is accepted in the project.

Chapter 4: Controlling and monitoring

Section 4.1: Sustaining progress

Unlike the previous three chapters and the remaining chapter on closing, monitoring and controlling is not a set of sequential tasks. The strategies laid out below can occur during any stage of a project. Since the bulk of work in the project occurs in the execution phase, most monitoring and controlling occurs in that phase. But the collection of data, measuring results, comparisons with project plans, and evaluating performance is just as important during initiation or closure. Project managers monitor and interpret data points during the project in order to control work flow and sustain progress to the project's end.

Gathering and digesting this data is a continuous endeavor that begins with the project and continues all the way through. As the project goes on, this monitoring information is used to ask and answer the following questions:

> ➢ Does the schedule help or hinder project progress? Is it realistic? Does it take advantage of positive team chemistries by ensuring complementary team-members share hours?
> ➢ Does each task have the most qualified people attached to it? Are they performing well?
> ➢ Does everyone on the team understand how they fit together?
> ➢ Does the budgetary plan need to be updated? Are there reasons to worry about higher prices?

Project controls regulate the collection and interpretation of the answers to these questions. Controls represent the processes by which project progress is regulated and accepted. The reporting requirements and frequency of progress reporting is a component of project controls. Controls are chiefly concerned with ensuring plans are being followed and finding any differences between the project plan and its execution. Controls establish the borders of permissible project work. A control is instituted to prevent a project problem from spreading. The **quality control plan**, completed during the planning phase of the project, specifies what acceptable and unacceptable standards for project work are. Project controls are therefore tied to evaluating project progress according to the quality control plan. If needed, controls will also lead to corrective action. Controlling for quality makes sure quality standards have been met.

Project monitoring deals with day-to-day managing and figuring out underlying strengths and weaknesses as the project plan is executed. Monitoring is not meant to discover problems, which is the objective of project controls. Rather, the goals of monitoring have to do with validating the scope of the project and ensuring every task is being completed. Monitoring is also responsible for finding the origins of reasons for tasks that are not complete, or to maximize new efficiencies. Looking for origins, instead of being distracted by small-matter symptoms, is key to effective monitoring. Like controls, monitoring is meant to improve the quality of project work. Controlling is active and forward-looking, while monitoring is meant to be investigative and looks at past performance. You can only have great controlling with great monitoring.

Monitoring project progress relative to the project schedule is a central component of monitoring. Having an easily-accessible accounting of the total amount of buffer the project has, and how much was used, is key. A project manager may not want to share the information on total buffer with every team member if they would like to be safe when it comes to deadlines and to be early. The details of the critical chain method and buffers are covered in section 2.3. But knowing the critical paths and amount of buffer continuously as project tasks are being completed prevents delays. If buffer is running low in monitoring reports, it may be necessary to add more resources to the project to increase the amount in reserve. When there are delays on paths that are not critical, the project manager should look weeks forward and see if the work flow in the non-critical area could affect a critical path down the road. With everything that deals with scheduling, monitoring should prioritize speedy reports and trigger the appropriate controls immediately.

Reserve analysis is the monitoring process used to assess resource reserves and decide whether and which additional resources are needed, or whether some resources are inefficient and not needed. Risk management processes, discussed in chapter three, are closely connected to reserve analysis.

Controlling and monitoring rely on updates on productivity and problems. **Productivity updates** specify the rate of task-completion per resource input. Productivity information can allow a project manager to predict progress forward and extend or collapse deadlines. Additionally, looking at resource productivity can allow for lagging team-members to be demoted or overachieving ones to be promoted. **Problem updates** provide ongoing information on identified project shortcomings. In larger organizations, team leaders handle problem updates. Whoever is responsible, the project manager must consider problem updates and decide if instituting additional controls are necessary to ensure the problem is contained. Both productivity and problem updates allow for project progress to be monitored day-to-day. A **corrective action** is taken whenever there is a problem with the execution of a project plan and project work is changed to correct the problem.

Keeping track of **intermediate steps,** not major milestones but rather day-to-day activities, helps a project manager keep track of progress through the life of the project and keep the project on schedule. Successfully completing intermediate steps on time and in budget will translate into larger success. Receiving feedback on smaller steps will help the project team accept criticism and respond to encouragement later on. A focus on intermediate steps will also prevent the most common issue that a project faces: a back loaded or delayed work schedule. It is routine, and a sad commentary on many project managers' efficacy, to see a project team frantically work until minutes before a deadline. Prioritizing intermediate steps will help ensure a timely completion of milestones instead of a scramble to complete them towards the end of the project. Frequent evaluation of the team, and keeping track of multiple goals, will amount to steps in the right direction. These intermediate actions translate into a successful project.

The monitoring and controlling processes come down to ways to increase accountability in the project. Team-members must be evaluated to know if the quality of their work is trending in the right direction or not. Frequent feedback provides increased accountability for past work. A project without accountability will compound past mistakes and the remaining time will be replete with even larger mistakes. A **forward view** is a control process by which a project manager lets the project team know what metrics they will be checking to measure progress the next time the project is reviewed. If the project manager meets with team-members the first week of each month, the next month will be discussed along with the past month during reviews. Accountability standards must be widely understood to ensure compliance.

Forward view thinking should inform **performance appraisals**. Performance appraisals encompass the many forms reviews of project work can take, including formal and informal methods. Work quality and efficiency, as measured in time and cost, are the usual metrics appraisals include. Performance appraisals produce reports that are used by the project manager, stakeholders, and higher management to raise awareness of issues and decide whether project plans need to be altered.

Many contracts require regular appraisals and keeping tabs on project progress. Time logs break each project team-members' day into segments tied to the specific work package. In the planning phase, the log's intervals should be decided upon (fifteen minutes, half-hour, hourly, etc.). Time logs can be audited by a contracting authority, especially in a dispute. Accuracy is critical to effectively manage resource reserves. Project managers should not tolerate guesses or round numbers to get to a required total amount of hours. A well-planned and efficiently-run project would not need to round hours up. To prevent such false reporting, be realistic about schedules and include times for team-members to work on tasks not tied to the project. Distinguishing between times spent on project and non-project work allows an organization to formulate better plans the next time a project is taken on.

Project appraisals and other forms of monitoring are accomplished by a wide array of information communication technologies (ICT) today. A quick web search will reveal the many options that exist for **integrated project management software**. The hallmark of these programs, almost all of which are online now, is a central hub to list and assign tasks to tagged team-members, provide comments and feedback, allow work progress monitoring and feedback, and usually some integration with word processing, databases, and other programs via attachments or inbuilt programs. The astonishing speed of change in ICT, driven by the insights of Moore's law of increasing computing power and global competition online, means these options will continue to increase.

As with most forms of technology, employing an integrated program management program can do more harm than good if done so with little preparation or experience. Any program used to track data and work performance should be shared. Team-members should make it a routine to look for feedback in the same place. Being able to compare past performance with planned performance should be the top priority when using any program. Quality control plans must define the time and resources allocated for software and online program introduction training.

You do not want to institute the use of project software without training in its uses and having a fluid handle on its workings. An even worse development would be to switch software during the second half of a project. Transferring data and the fixed transactions costs of switching will likely be greater than any benefit. In general, a good rule of thumb is to see what software your project team does best with and move in a direction that builds on existing organizational strengths. A great teaching tool is screen-sharing video. For new software, there could be recorded and accessible screen-shares that show project teams prioritized functions in chosen software.

Section 4.2: Controlling project changes

Project plans are rarely executed with complete fidelity. All authorized changes to a project plan are **scope changes**. On the other hand, **scope creep** comprises all changes to the project plan that are not authorized as they happen. All extra work, no matter how small an addition to the work breakdown structure, constitutes scope creep. When a small extra task here or there adds up for the course of a project, the amount of excess work will creep up on organizations and quickly bring about unexpectedly large bills. The purpose of project controls is to prevent or at least contain scope creep. **Reworks**, which are improvements to subpar products and project outcomes generally, are another type of scope change. Reworks require testing for quality and ensuring the intervention is reasonable and does not go beyond the required changes.

Details are the enemy of scope creep. The greatest defense is to clearly lay out each project task in deep detail and try to not deviate from the plan. Clients often ask for extra work during the project. These alterations can be avoided by clearly setting project limits and wisely using resource reserves. Game-planning to better understand the consequences of a small upcoming change on tasks further along in the project will help contain scope creep. As always, communication is key to effective project management. Letting clients know whether or not their extra request can be accommodated, and being willing to say no without more resources, leads to a better outcome.

The **issue log** (or change log) handles all extra work and attaches owners and due dates to each new project task. A growing issue log is a sign of scope creep. Issue logs identify the current obstacles preventing project progress. Team and stakeholder meetings should review issue logs to get everyone focused on the same matters. Analyzing the impacts of each ongoing issue will prevent being blindsided by an unexpected consequence. In larger organizations, a **change control board (CCB)** has the responsibility to approve requested changes and set the cost and timeline for an additional change. The board's rules should govern how change requests are submitted, and what the turnaround time is for a decision.

If mandated in your project's requirements, a CCB's central focus demonstrates the close relationship between change and control. In arrangements with a CCB, a project manager's workload is lightened as the board will make the call on whether to reject or accept a request. In some instances, a project manager will serve on the CCB as well as informing it. CCBs have specific methods for dealing with emergency requests close to deadlines. Specialized boards play important roles in controlling and monitoring.

The process by which project issues are requested and documented should be controlled. As always, a best practice is to have requests for a project change submitted in writing. The exact request should be detailed. The requester should vet the final result to ensure it can solve the original issue. The consequences of a change in one area on all other project milestones should be evaluated and studied. The reasons for accepting or rejecting a requested change should be spelled out. Team-members in every project area that will be changed should be brought into the process of updating the issue log and coming up with an action plan. The detailed steps to implement the change should be detailed and each should have their own timelines.

A special type of control processes is **configuration management**. Configuration management controls and standardizes the rollout of approved changes to the activities and goals of the project. Configuration management focuses on consistency in whichever area in which it is used. New project sequences, deliverables, or changes to the capabilities of products are common types of configuration changes. If an engineering firm is producing a carbon fiber for a client in the aerospace industry, configuration management protocols would oversee changes to the product, such as adding heat resistance. Unlike standard changes, changes to configuration require more resources and possibly expertise. Changes to project baselines and documents also constitute changes to configuration.

Any change to configuration should be analyzed in terms of risk, the issue log, and team morale. Risk reviews are updates to the risk management plans. In the wake of any configuration change, risk reviews should be updated and checked for any missed dates or errors. Issue log entries should be reviewed to see if there are any current project problems that would be affected by the configuration change in question. Often, changes to one aspect of the project will ignore similar project issues that need to be addressed. Opportunities to address issues with one fix should be sought out. Team morale could be the deciding factor in determining project success. Changes to the project scope should be explained and sold to the team. Project controls will not work without team buy-in.

For substantial changes to project plans, or whenever a new project plan is adopted, **rebaselining** is necessary. The new baseline will determine remaining project tasks and offer new quality checks. Any time a change occurs to critical path project activities, rebaselining will be needed as critical path changes will, by definition, delay the entire project. Schedule changes and any delays should be explained to all stakeholders. If delays are not communicated and approach unexpectedly, it could cast a shadow on other successful parts of the projects. It is better to be transparent when it comes to delays as a longer timeline will affect everyone involved.

The decision to rebaseline should not be taken lightly. A project manager should clear the new set of guidelines with project stakeholders and get their consent. Announce the new baseline expectation to all project partners, audiences, and subcontractors and take the time to answer any resulting questions about the changes. A new baseline in one area, like a budget baseline, must be reflected in changes to all other project plans. Much of the work with rebaselining is taken up with recalibrating every project timeline, schedule, and budget.

Forward projections are used many times over the course of a project, but are essential when controlling project execution and when rebaselining is necessary. Forward projections use past project data, such as productivity and how far from reality project plans proved to be, to predict the future course of the project and the effects of all the changes that go into a new baseline. **Econometric methods** are used to identify variables that might change your project's forward projections. Methods include statistical ones like regression analysis and autoregressive moving average (ARMA). Monte Carlo random sampling methods are a big part of econometrics.

Many controlling processes can only begin in the latter stages of the project. Especially in a project that produces a product, there needs to be a finished project that can be evaluated. A **first unit** is the first complete product the project produces before producing many more of them. More than any other unit of further production, it is tested for defects. In a well-controlled project, all subsequent products will be of higher quality and require less total work.

Sampling can be done for all projects and is another method to control for quality. Sampling means taking a fraction of the total to draw conclusions about the total. Sampling can be made more precise by correcting for over-representation and aiming to ensure the subset of samples taken are representative of the entire range of possibilities. **Variable sampling** measures and compares quality on a scale to measure the level of difference from product to product. A hard-and-fast rule in sampling is the **rule of seven**, which states that seven examples out of a large set of observations are needed to make a real trend and ensure you are not falling for anecdotal or misrepresentative evidence.

Section 4.3: Funding and project progress

More often than not, a project requires more or fewer funds to complete than what was planned for. Any work outside of the scope, team-members proving less productive, changes in resource prices, poor subcontractor coordination, transaction mishaps like late orders, and changes in organizational structure are common causes of cost variance. The concepts in this section will help a project manager navigate through these issues while controlling a disciplined budget. Monitoring the project's planned budget requires a specific set of skills compared to a personal account. The costs in a budget are dynamic, changing over the course of a project, and require constant monitoring. The **burn rate** indicates the speed with which project funds and resources are spent. All projects have a burn rate. They key is to have a burn rate that matches budget plans.

Forecasting future costs is often difficult and leads to a changing burn rate as the project matures. **Overhead costs** represent one area that must remain constant for budget plans to remain accurate. Overhead costs are those things that are not tied to specific project work, but instead cover basic operations. Team-members' health insurance, office supplies, renting office space, and utility costs are examples of overhead. **Administrative costs** represent the cost of paying project team-members' salary, servicing contract fees, and general costs that allow an organization to operate at capacity. Administrative costs are tied to project work, while overhead is not. For example, if you add one more person to your project team's existing office, administrative costs will go up, while overhead will not. If your overhead or administrative costs increase mid-project, your burn rate will increase.

There are many techniques that can be used to increasing project cost savings. **Project life-cycle costing** compiles all costs every project task will accumulate over its lifespan. It groups categorical costs, such as acquisitions, operations, and disposals, and compares alternative work flows to find savings. **Value engineering** optimizes project performance and cost by primarily looking to eliminate unnecessary costs. A higher than expected burn rate will lead to value engineering to analyze costs.

The need for more resources is a sign that a project burn rate is greater than planned. As with all requests for change in a project, funding requests should be submitted in writing. Budget requests should estimate the cost of the requested change. If it is a large request, the written request should include the range of possible costs from an upper to lower limit. Depending on the organizational setup, the requested change could involve a procurement department, multiple vendors, subcontractors, and legal expertise if a contract needs to be modified. This complexity, along with the dynamic nature of upcoming project costs, requires the project manager to keep track of a great number of budget data points.

Earned Value Management (EVM), or Earned Value Analysis, allows for a project manager to remain current with budgetary and spending plans in detail. EVM reviews a project's cost status relative to schedule status based on expenditure alone. Value of work is understood to coincide with the amount budgeted to complete the work. If two-thirds of the budget has been used, but the project is only half completed, EVM will provide insights into where there was overspending and funds went. For tasks not yet completed, EVM is an effective method to anticipate future problems and find out where spending does not produce as much as expected. The spent budget should be categorized by contracts, accounts payable, unplanned expenditure, and so on, to make trends clear and find hidden savings. EVM is how monitoring and controlling looks when it comes to the project funding. The techniques described below can be understood as components of EVM.

For the purposes of EVM techniques, there are three types of activities to measure work performance. **Apportioned effort** is project work that is understood to be continuous and not divided into discrete work packages. By contrast, **discrete effort** is easily measured project work that produces a specific output. **Level of effort** is project work measured in time and is not necessarily tied to a specific deliverable. These measurements are integrated into the metrics below.

For successful EVM, data needs to be collected on project progress relative to the plans. Many of these terms are known by two names, and both names will be indicated the first time these values are referred to in what follows. **Budget at completion (BAC)** is the total budget allocated for the project in project plans; the sum of the budget for each project phase. BAC is usually plotted over time, and changes with requests for increased spending. **Planned value (PV)** or **budgeted cost of work scheduled (BCWS)** is the planned budget for work in a specific date range. **Actual cost (AC)** or **actual cost of work performed (ACWP)** is the real amount spent for the work done in a specific date range. **Earned value (EV)** or **budgeted cost of work performed (BCWP)** is what the plan allocated for the actual work completed. EV determines whether the project is over- or under-budget by calculating what should have been spent in a specific date range. The equation for EV is:

$$EV = BCWP = (BAC)(percent\ of\ project\ actually\ completed)$$

To explain EV in greater detail, let us stipulate a BAC of $100. If the project is a quarter completed, the EV is $25. If the AC exceeded $25, the project is over-budget. If AC is less than $25, the project is under-budget. While indicators like EV are usually expressed in currency, they can also be expressed as hours or percentages. While the examples examined here are calculated in dollars, the same rules apply if using hours or percentages as the metric.

As the equation for EV shows, it requires the project manager to determine the "percent of project actually completed." Even for a highly organized and regulated project, meticulously taking note of each quarter-hour spent on a task could prove to be difficult. There are three methods to resolve this issue. The **percent-complete technique** estimates the exact amount of the project completed. This method requires the most meticulous project monitoring to not lead to an erroneous EV estimate. The percent-complete technique is often viewed as depending on informal judgments of project progress, and not attainable. By contrast, the **milestone technique** sacrifices preciseness for an either/or approach. The milestone technique measures project progress by the total amount of projective activities that are complete. If a project activity is partially complete, it does not count towards the percent of project actually completed calculation. For firms using the milestone technique, EV is calculated infrequently. If it is calculated while project work is underway on many tasks that have yet to be completed, it will seem like the project is over budget in EVM calculations. The **fifty-fifty technique** compromises between these two methods. With fifty-fifty, a project activity is zero percent before relevant work begins, fifty percent while work is ongoing, and one hundred percent complete once work is finished. The total amount of the project completed is derived from each project activity's progress.

The schedule is an integral factor in EVM-driven expectations. The following EVM indicators, focusing on both schedule and cost, are interconnected and build on the concepts of earned value and actual cost. Some of the following terms appeared in chapter two on planning. Their meaning is unchanged, but here you will get a better sense of their connections and the relevant equations.

➤ **Schedule variance (SV)** calculates the difference between a planned budget for a portion of the project and the actual amount spent on the same portion of the project. The SV can determine the extent to which you are behind, ahead, or up-to-date with the schedule by showing the gap between expected and actual completion. The difference between Earned Value (EV) and Planned Value (PV) provides SV:

$$SV = EV - PV$$

➤ **Schedule performance index (SPI)** calculates the quotient of earned value (EV) to planned value (PV). SP is an easy way to see if the project is ahead of schedule or behind schedule and how efficient project work to a certain point has been. Another benefit of SPI calculations is that it reveals how much of the planned schedule was actually accomplished. If SPI is calculated to be above one, the project is ahead of schedule. If SPI is calculated to be below one, the project is behind schedule. If SPI is equal to one, the project is right on schedule. The equation for SPI is:

$$SPI = \frac{EV}{PV}$$

➤ **Cost variance (CV)** calculates the difference between the earned value (EV) and actual cost (AC). Allowing a project manager to quickly see the amount they are under or over budget, CV clearly shows the magnitude by which cost plans have or have not gone awry:

$$CV = EV - AC$$

➤ **Cost performance index (CPI)** calculates the quotient of earned value (EV) to actual cost (AC). The CPI indicates the project's cost-efficiency: how much project work is completed for the expenditure. The CPI can help prevent unexpected over-spending. If CPI is calculated to be above one, the project is under-budget. If CPI is calculated to be below one, the project is over-budget. If CPI is equal to one, the project's planned budget is being followed to the penny. The equation for CPI is:

$$CPI = \frac{EV}{AC}$$

Both earned value management indexes can be made into cumulative versions. If, for example, CPI is used multiple times during the project's execution, a project manager would be wise to calculate **cumulative CPI** to analyze trends. To calculate cumulative CPI, the cumulative sums of EV and AC must be found and divided. It is similar to the non-cumulative equation. Non-cumulative CPI analyzes a discrete time period, while cumulative CPI analyzes a larger period, and is used especially at the end of the project. Actual cost (AC), planned value

(PV), and the variances based on them are definitionally concerned with smaller time periods and do not benefit from a large number being compared together. **Cumulative SPI** is similarly used and calculated for the same reasons:

$$Cumulative\ CPI = \frac{Cumulative\ EV}{Cumulative\ AC}$$

$$Cumulative\ SPI = \frac{Cumulative\ EV}{Cumulative\ PV}$$

There is another index used to calculate project efficiency and predictions based on existing trends. The **to-complete performance index (TCPI)** determines the cost performance needed to meet the project's planned budgets, or the budget at completion (BAC). TCPI looks at remaining project tasks and comes up with the amount these tasks must come in under at to keep the project within budget's confines. It does this by examining the difference between earned value (EV) and actual cost (AC) relative to BAC. The equation for TCPI is:

$$TCPI = \frac{BAC - EV}{BAC - AC}$$

Everything covered in this chapter section, Funding and Project Progress, can be understood as improving **project forecasting**. Forecasting uses past project information to predict future performance, conditions, and task completion timelines. Forecasts take work efficiency data, and how well executed plans held up, into account. Forecasting leads to two more significant earned value management indicators. **Estimate at completion (EAC)** is a forecast of the total cost of the project. EAC can be calculated at different points of the project with the equation:

$$EAC = AC + ETC$$

Actual total costs (ATC) is the exact amount spent at a particular point of the project. Before the project begins, AC is zero. The sum of AC and the **estimate to complete (ETC)** the rest of the project provides EAC. Simply put, ETC is a forecast of the amount of money needed to finish the remaining parts of the budget after project spending has begun. The equation for ETC is a version of EAC and shows earned value (EV) plays a central role in forecasting:

$$ETC = EAC - AC = BAC - EV$$

As a reminder BAC, budget at completion, is the total budget allocated for the project in project plans; the sum of the planned value (PV) for each project phase. Earned value (EV) is what the plan allocated for the actual work compared to the relevant point in project work. Therefore, ETC forecasts are useful in deciding how well project spending plans have held up as the project is being executed.

There are many variations on these versions of EAC and ETC, and they are used in many specific instances. Be sure to pay attention to the mathematical law of transitivity. In the ETC equation, for example, ETC can be replaced with BAC-EV. EAC can also be calculated with the CPI (cost performance index) understanding of efficiency. This type of EAC is termed the **EAC forecast for work performed at present CPI**. EAC in this sense is tied to calculations of work performance productivity and has a simple equation that is a result of substitutions in the EAC and CPI equations:

$$EAC_{CPI} = \frac{BAC}{CPI}$$

Projects often run into a sudden event that increases costs. When these events are reasonably expected not to recur, EAC is calculated as one-time cost variance EAC. In the following equation, AC accounts for the unplanned, nonrecurring project cost:

$$EAC_{CV} = AC + BAC - EV$$

On the other hand, if a significant cost-variance is not a one-time event but likely to repeat, another formula must be used. EAC in this instance is termed: repeated cost variance EAC. The CPI accounts for the multiple, unplanned cost increased in the following equation:

$$EAC_{RCV} = \frac{BAC}{CPI}$$

The most complex version of EAC is **EAC forecast for ETC work considering both SPI and CPI**. Managers only employ this type of EAC when the cost performance is inefficient and project deadlines cannot be met. The equation focuses on the combined effect of SPI and CPI trends on the estimated cost to complete (ETC) the rest of the project:

$$EAC_{SPI \& CPI} = AC + \frac{ETC}{(CPI)(SPI)}$$

At the end of a project, EAC can be used to calculate **variance at completion (VAC)**. VAC produced the difference between the planned budget and EAC:

$$VAC = BAC - EAC$$

Section 4.4: Practice exam questions on controlling and monitoring

1. Which of the following is an example of a corrective action?

    a. After an unexpected delay in project work, the project manager alters team composition to hasten project completion
    b. After a breakdown in team communication, the project manager requests that all communication follow a new protocol
    c. After an unexpected delay in project work, the project manager requests and evaluates schedule change requests
    d. All of the above

2. Which of the following is the method for determining the schedule performance index (SPI) of a project?

    a. Find the quotient of earned value and planned value
    b. Subtract planned value from earned value
    c. Find the quotient of earned value and actual cost
    d. Subtract earned actual value from budget at completion

3. You are a project manager for Enlighten Partners, a consultancy firm that specializes in online marketing. During a project for a new client, a new technique is implemented that immediately increases sales. The client wants development of this technique to be more prominent in upcoming project work. The project scope, however, did not account for more work on this new technique. How is the change control board involved in handling the client's request?

    a. The change control board deals with internal issues and does not decide issues dealing with clients
    b. The change control board is responsible for reworking the relevant project plan and timeline
    c. The change control board is tasked with reworking budget agreements to accommodate the client's request
    d. The change control board approves or denies all requests

4. What is the purpose of a configuration management system?

   a. It allows a project manager to remain current with budgetary and spending plans
   b. It allows a project manager to prevent all unauthorized changes to the project plan
   c. It allows a project manager to specify acceptable and unacceptable quality for work
   d. It allows a project manager to coordinate the uniform rollout of approved changes to the activities and goals of the project

5. What is the difference between controlling project quality and validating project scope?

   a. Controlling project quality leads to accepting work results, while validating project scope sees the project manager ensure that work processes have not veered from project plans
   b. Controlling project quality ensures work results satisfy work standards, while validating project scope leads to accepting work results
   c. Controlling project quality ensures work processes have not veered from project plans, while validating project scope sees the project manager accept work results
   d. Controlling project quality assesses resource reserves and decides whether and which additional resources are needed, while validating project scope sees the project manager ensure work processes have not veered from project plans

6. A project that is tasked with creating a new patient-doctor portal has suffered delays and come in over budget. A project manager has all the required data to hone in on the project's issues in order to address them. Which of the following is an incorrect way to analyze the project's flaws?

   a. Using the issue log to determine a new baseline with updated project tasks and new quality checks
   b. Examining the schedule performance index by finding the quotient of earned value and planned value
   c. Examining the cumulative schedule performance index by finding the quotient earned value and actual cost
   d. Examining the estimate at completion by finding the sum of actual cost and the estimate to complete

7. In the middle of project work, a new stakeholder request triggers a change to the work breakdown structure. The change does not alter project activities on a critical path. Such a change necessarily leads to which of the following?

    a. Budget changes
    b. Schedule changes
    c. Scope changes
    d. None of the above

8. What distinguishes project control from project monitoring and what do they have in common?

    a. Project monitoring deals with day-to-day evaluations, project controls regulate project progress, and both occur during the execution phase of a project though the phase overlaps with other phases
    b. Project monitoring regulates project progress, project controls maintain quality, and both occur during the execution phase of a project
    c. Project monitoring regulates project progress, project controls maintain quality, and they are together their own project phase
    d. Project monitoring deals with day-to-day evaluations, project controls regulate project progress, and they are together their own project phase though the phase overlaps with other phases

9. Why is negotiation the preferred method of resolving disputes in the controlling and monitoring phase of a project?

    a. When getting project stakeholders to first agree to project goals, negotiation is utilized to reach a feasible compromise
    b. If there is a dispute among project contractors, negotiation allows for the settling of differences and leads to a verdict to be issued regarding obligations agreed to in contracts
    c. As project activities are being regulated for quality, negotiation is preferred to smooth over differences and continue project work
    d. As project activities are finished, negotiation ensures that team-members stay on the same page and remain dedicated to the project until all work is completed

10. At a certain point in project work concerning the total project and one deliverable, the following is true:

- The budget at completion is $1000
- The planned value is $200
- The actual cost is $195
- The earned value is $250

Based on that information, which of the following is correct?

a. The cost variance is -$55
b. The cost performance index is 1.5
c. The schedule variance is -$50
d. The to-complete performance index is .9

Section 4.5: Practice exam questions on controlling and monitoring – answers

1. Answer: D. A corrective action is taken whenever there is a problem with the execution of a project plan and project work is changed in order to correct the problem. All of the options introduce a new rule to get things back on track. A preventive action deals with limiting the impact of a risk event and is often confused with a corrective action.

2. Answer: A. The SPI is found by the formula in which earned value is the numerator and planned value is the denominator. Answer B is schedule variance. Answer C is cost performance index. Answer D is variance at completion.

3. Answer: D. There is a lot of extra information in this question meant to distract. It does not matter who makes the request. If a project has a change control board, that entity is responsible to either accept or deny a request. All of the particulars, like budgets and timelines, are not its purpose. These other considerations could, however, shape its decision.

4. Answer: D. Configuration management systems focus on putting approved project changes into place. Answer A describes earned value management (EVM). Answer B describes scope creep. Answer C describes a quality control plan.

5. Answer: B. Only answer B has the correct combination. Scope creep, preventing unauthorized changes to the project plan, reserve analysis, assessing resource reserves and deciding whether and which additional resources are needed, appear in the other answer choices.

6. Answer: C. Only answer C has the incorrect association or equation.

7. Answer: C. A change to the work breakdown structure is automatically also a change to project scope. If the change were on a critical path, then the project schedule would need to be updated. Although the budget usually is altered with any scope change, the question did not provide sufficient information to determine cost changes. Scope change can cancel one cost and replace it with another so there is not net budget change.

8. Answer: D. "Controlling and monitoring" is a separate phase in project work. The processes and techniques it groups together can be used in any phase in the project. Project monitoring deals with day-to-day managing and figuring out underlying strengths and weaknesses as the project plan is executed. Controls are chiefly concerned with ensuring plans are being followed and finding any differences between the project plan and its execution. Controls establish the borders of permissible project work.

9. Answer: C. The key to answering this question correctly is finding an example that illustrates negotiation and occurs within the controlling and monitoring phase. Answer A is a good use of negotiation, but belongs in either the initiation or planning phase. Answer B could occur in the correct phase but it is an example of arbitration, not negotiation. Answer D occurs in the closing phase and is an example of collaboration more than negotiation. Only answer C occurs within the controlling and monitoring phase and shows key components of negotiation strategies.

10. Answer: D. The question requires knowing and correctly working the relevant equations. For the to-complete performance index (TCPI), the following work must be done:

$$TCPI = \frac{BAC - EV}{BAC - AC} = \frac{1000 - 250}{1000 - 195} = \frac{750}{805} = .9$$

The correct cost variance is $55, the correct cost performance index is 1.9, and the correct schedule variance is $50 (not negative).

Chapter 5: Closing

Section 5.1: Right before the last deliverable

All projects have a completion date, the point in which all project activities are done and delivered. Although the ways a project can end are as varied as the shapes projects can take, there must be an ending for a set of activities to constitute a project. Many projects taper off slowly, with a few activities ongoing after the vast majority of work has been completed. Other work assignments may not have a distinct end like a project. This second to last section deals with the project right before it closes; the next section focuses on closure and the importance of the project for the project manager's organization.

Of all the plans made before project work began, the **project closure plans** are likely the only plans to have remained unaltered through the execution phase. The sooner a project manager can update closure plans, and recalibrate them with changes made to other project plans, the better. Before completion tasks begin, it is necessary that the project team knows what updated closure activities are. Ideally, closure plans' activities and timelines will be continuously updated as the project evolves.

Maintaining a **project checklist** is an effective way to keep track of all unfinished and remaining project activities. The activities that are not checked off will make up a good portion of project closure activities, even if these activities are holdovers from an earlier phase of the project. When everything is checked off, the project manager can verify that each deliverable is complete and can close the project. Accomplished professionals in a variety of fields attest to the power of checklists in avoiding common mistakes and oversights. Surveying project stakeholders towards the end of the project is an effective tool to guard against forgetting something and to make sure project work continues to revolve around core stakeholders.

Complementing checklists, keeping up regular team meetings and expectations through the closing phase will ensure effort and productivity will not taper off. It is likely that team-members or contractors will leave project work as their expertise was only needed for earlier project activities. As long as the project manager oversees the work of someone else, the total team should be called together and understand the importance of finishing project work strong. A sign of a strong finish is the final dozen work packages not taking longer than prior sets of a dozen work packages. Ideally, the last portion of work will be completed quicker with improved productivity as long as the project team remains intact. Stagnant or regressed productivity by team-members during the project's close is a worrying sign, and may detract from a project manager's otherwise good work.

Checklists and efforts to sustain team morale and productivity are attempts to avoid some of the most common problems as a project ends. Here are the usual explanations for difficulties at a project's end:

> *Confusion about remaining activities.* If the project plans were vague about the necessary steps to finish the project, it will be difficult to finish strong. The earlier a project manager can detail all the activities needed in closing, the better. While work in the execution and controlling and monitoring phases may overlap, closing activities are generally more distinct.
> *Waning team effort.* No matter if the project ends because everything went to plan or things went disastrously and it was cut short, most stress will be in the past as closing activities begin. A project manager should step up efforts to keep team-members sharp and engaged as the project progresses, and be most proactive as the project ends. Team chemistry may also be affected by some team-members being reassigned and possibly even acquiring new team-members. Team building should be a focus from the first to last day of the project.

There are significant benefits at a project's end as well. As covered earlier, the S price model curve stipulates that costs are significantly lower at the end of a project. Hopefully, budgetary issues will not be an issue in the closing phase. No matter how a project ends, the majority of work will be done and most deadlines have passed. Closing features fewer moving parts and a lower chance of risk events occurring. But the level of difficulty varies by the type of project closure. Ranked from most straightforward to most difficult, there four models for project closing: extinction, addition, integration, and starvation.

Extinction is the ideal type of project closure. Extinction implies a project was successfully completed and accepted. An extinct project does not linger on and has a definitive endpoint. Stakeholders were satisfied with the results. Most project lessons and features are offered with an extinction-type project in mind. A project with an extinction ending does not imply that everything went according to plan. The extinction date may have been delayed, or the budget exploded. But when closure happens, an extinction closure is final.

The next two types of closures may also be successes, but do not have clean endings. Both addition and integration project closings were likely unexpected when the project began. Some project stakeholders may have preferred an extinction ending.

Addition is when a project becomes a permanent feature of the firm, such as evolving into a separate unit within the organization with ongoing operations.

Addition necessitates that more resources are added to the project to make it permanent. For example, if a team tasked with improving technology competency executed their project so well that the school district begins a permanent effort to increase technological fluency, that project ended by addition.

Integration ends one project by merging it with another project, or folding it into an existing work unit. Projects that end in this way see their resources (such as equipment) and team-members either assigned to different projects or returned to their prior roles. Whereas addition changes the capabilities of the performing organization by adding a dedicated unit concerned with a new type of capability, integration does not add a new capability to the organization. Integration does not have to be a negative outcome. A firm may have two ongoing projects for the same client, and that client may want a united project that includes greater resources. The project that is integrated into another is said to end even if the overarching project has a lot of life left.

Starvation ends a project by withdrawing resources until nothing remains. A decision to starve a project implies failure and an unplanned early end. Since even in the most drastic situations work must taper off in some form, even rapidly ended projects are starved. Starvation can extend for a lengthy period if large deliverables are canceled and no longer a goal. The difference between integration and starvation is that integration means the project work will continue on in some form, while starvation implies the firm moving on from the project's specificities full stop.

A negative **viability decision** leads to starvation. After a series of work plans going awry, deadlines missed, budget blow-ups, or other mishaps, viability decisions decide whether to close a project early. A project sponsor and the larger organization make the decision, and it is up to the project manager to organize and execute the starvation closing plan. A project can be cut off early for many reasons. Losing an important team-member, not being able to access an important resource, the firm losing money in another area and having to scale back, or the client reducing the project's scope are common factors. Schedule and budget issues are usually symptoms of one of these underlying causes.

When projects end via extinction, addition, or integration, standard closure practices must be followed. Starvation implies an accelerated timeline, and depending on how much project work was achieved, some of the following may not be applicable. There are many involved processes in standard closure practices that may be completed in the most efficient order. Attaining buy-in for the order chosen from the remaining project team is more important than chosen order. As with the project as a whole, commit to a plan and only deviate from it by following a standardized procedure.

Handover protocol ensures that all necessary project products and artifacts have been successfully signed off on and that documents are archived correctly. This includes ensuring handover documentation has been successfully delivered or stored. Handover protocol also covers the logistics of product completion, making sure all users of a project's output are able to use it successfully. **Acceptance criteria** details what constitutes project completion per project charter rules and how all parties to the contract express their agreement that the project is complete. Closing out the procurements by ending contracts for project services is also a process that will occur once a project ends. Finally, team-members should be promptly released and reassigned once the project is closed.

The procurements process must be finalized with contracts settled, and all parties to a contract must agree to the project's end. In a starvation-ending, this could prove to be a complex affair. If a starved project has outstanding contracts that are no longer needed or can no longer be fulfilled, **arbitration** may be needed if contracts allow for that. Arbitration sees an independent third-party come in, hear sides of a dispute, and make a legally binding decision. If arbitration is not possible, a civil court may settle contract-related disputes. A prior agreement about an **indemnity** may obviate the need for arbitration or a trial. An indemnity protects parties against future loss by requiring reimbursement in the event of cancellation. Indemnities can take different forms. **Ascertained or liquidated damages** is a set amount, either fixed daily or on some other timeline, to be repaid in the event of a breached contract.

Whichever way a project is completed, the **validate scope** process must be completed. Validating scope reviews project documentation for accuracy and completion and validates each deliverable as complete. Even for a project that is terminated early on, the scope of completed activities should be validated to ensure proper billing, and so mistakes are not repeated.

The form a project closing takes shapes work on the last deliverable and determines closing processes. The following section provides insights into all project closing types.

Section 5.2: Closing the project

Project completion is not the same as project closure. After the final delivery stage is carried out and the closure processes outlined in the last section are checked off, all project activities are complete. But between project completion and the closure date, all project work must be evaluated so lessons are learned and future projects can go smoother. It is after every project activity is completed that the project manager can assess the effectiveness of project plans and performance. Project closure should not begin until all project work is complete and all data needed is in hand. Additionally, effectively closing a project and reflecting is the stage at which most of a project manager's professional development takes place. The project is "open" until all evaluations are finished.

Some processes exist at the boundary between project completion and closure. As a project is completed, more and more of the project team will be released. This is often not a formal process, but depends on the needs of the firm and the size of the project. Even after reassignment, however, former project team-members should be available. When evaluating the course of the project, getting input and answers from involved team-members about critical moments in project work is necessary. A project manager should arrange for this communication to be possible, as well as having project stakeholders review evaluation documents and make their own contributions. This is in addition to a project manager's duties with former team-members, which includes making sure there is no downtime before reassignment occurs, and helping smooth out transitions.

Carefully completing intermediate phases of a project is the best preparation for project closure. After completing a large set of deliverables during project execution, a project manager should evaluate plans and performance during that phase in order to improve as the project evolves. The processes during project closure can be a combination of phase closing processes undertaken earlier in the process. Ending a phase involves many of the processes also seen when completing a project, including formal acceptance documents from clients and disseminating notices to the project team and project stakeholders about project evaluations. Finalizing documentation is part of completion. Evaluation processes are the heart of closing the project.

The documents that the project completion process produce inform project closure. Without accurate final documentation that allows for the original project scope to be compared to the total amount of project work, it is difficult to analyze the total project and draw from it useful lessons for the future. Final project documentation including billing and project teams' work performance and the productivity metrics tracked through project execution. The immediate goal of project completion is to ensure stakeholders and customers are happy.

But for project closing, the more relevant goal is to have submitted detailed records to facilitate a straightforward review.

A **lessons learned report** uses documented details to focus on corrections made during project work. Beyond narrating the process by which errors were made and plans altered and the results of those corrections, the report includes valuable insights learned during the project generally. These insights could include debriefs on what it was like to work with all the associated project contractors or effective communication protocols. All projects need a lessons learned report, even ones that are terminated early on.

Collecting and interpreting information takes up most of the time in describing project lessons. Expenditure details, scope changes, clients' positive and negative comments, other managers' positive and negative comments, and schedule performance should all be analyzed. Progress reports, project logs, budgets, schedules, and communication memos are the most common tools to narrow in on lessons. All lessons-learned reports should have:

> ➢ A review of original purposes of the project for all parties and the degree to which those expectations were met
> ➢ Analysis of the work breakdown structure's objectives and which, if any, were not met and why
> ➢ Input and evaluation from team-members and stakeholders
> ➢ Verbatim snips from project logs, communication, or surveys
> ➢ Final budgets

Another critical part of a lessons learned report is the **controls review** portion. The control review analyzes the strengths and weaknesses of how changes to the project plans were handled in the project. It also proposes ways to improve project controls. The authority a project manager had to execute changes and handle requests should also be addressed, especially if those controls played a part in missing deadlines.

Post-project evaluation, post-project reviews, and a project postmortem are other terms for lessons learned reports. Though the "postmortem" evokes negative images, it can be used for a project that ends successfully, as well as projects are starved off. All these reports share attributes and look at what went right and wrong during project work. They each produce a formal report with subheadings. The report should highlight project team achievements, especially when the team accommodated additional requests or handled a risk event in stride. Processes that led to increased productivity should be detailed step-by-step, such as the details of a new communication protocol. The report should contain suggestions on how to avoid failures the project experienced.

The final review is often disseminated to all parties associated with a project. In many contracts, it is a requirement that a lessons learned report be produced.

All parties to look forward to the lessons learned report to improve. There is almost always a meeting that the principals attend that sees the presentation of the lessons learned report. Taking the principals' perspective and speaking to the original hopes and drivers of the project will help the meeting go smoothly. Did the project change what it set out to change? The importance of the report and presentation should not be underestimated. Even years afterward, organizations can review these reports to decide whether they should do more business with one another.

The **procurement file** is the collection of all records dealing with contracts attached to the project. The file should be referenced when deciding with whom to share the lessons learned report and to ensure that each of these relationships is reviewed in the report. All after-project work cannot be done without complete information on full costs breakdowns and performance reports. Do not be shy about requesting additional information if a hole is found in the procurement file or any other report during evaluation.

Procurement closure is the process by which a contract is terminated, either successfully or unsuccessfully. Procurement closure can be done many times during a project, depending on how many contracts were associated with the project. **Administrative closure**, on the other hand, should only be done once when a project is closed out. Administrative closure signifies that all information needed regarding the project has been received and accepted by the appropriated parties and further project communication is not expected. Administrative closure is done only after post-project evaluations are completed.

All of the post-project evaluations are pursued to create an ethos of **mature project management** in your firm. Mature project management refers to the performing organization having a corporate culture, established methodology, and deep understanding of project-managing fundamentals that lead to successful projects. The more lessons learned reports an organization has, the more mature its project management expertise should be.

As with everything at the end of a project, success in final project activities will depend on systems put in place earlier. If project errors were handled quickly without rules for processing requests, it would be difficult to go back and think through all the steps taken to improve the situation. Lessons learned depend on a project manager holding the project team accountable when it comes to reporting requirements. Just because project closing is largely for internal consumption does not mean the plans for producing the closing reports should be skeletal. Planning out what is necessary to write in-depth closing reports must be done as rigorously as any other plan.

Thinking project closing activities are simple is the most common mistake during these final project stages. For example, tagging certain emails to be included in the lessons learned report as examples could be a step taken earlier on to strengthen closing activities. Project managing during the scope of one project is cumulative. Dropping the ball during an earlier stage will make subsequent work more difficult. Closing activities suffer the most from earlier mistakes. Involving the expert judgment of team-members and stakeholders as well as utilizing a project management information system (PMIS) helps forestall these issues.

Product verification is an ongoing process that is only completed when the project is closed. Product verification verifies that project work is indeed complete and matches contract requirements and/or stakeholders' expectations. If, for example, when reviewing a technical project, a capability is found to be an error, it should be corrected. A completion document with inaccurate information should likewise be corrected. A project manager should temporarily reopen completed project activities.

Reopening completed activities, especially when inconvenient, is a case of respecting a project manager's **code of ethics**. This code focuses on four areas that allow a project manager to do the right and most effective thing: responsibility, respect, fairness, and honesty. There is no clear line between doing what is efficient and ethical—the two reinforce one another. As a project is being closed out, this rule should be abundantly clear. Only a truthful reflection of what went right and what went wrong in a project will allow for a project manager to improve and not fail similarly during the next project. Openly incorporating lessons from failures will improve self-confidence and leadership abilities. More than anything else, an ethical management style will lead to better team dynamics. A project manager is only as good as the team they lead.

Section 5.3: Practice exam questions on closing

1. How does the closing phase fit within the framework of an entire project?

   a. Ideally, the closing phase, like the initiating phase, is near a tenth, the planning and controlling and monitoring phases are near a quarter, and the executing phase is near a third of total project time. The closing phase does not occur until the executing phase is complete.

   b. Ideally, the closing phase, like the initiating phase, is near a fifth, the planning and controlling and monitoring phases are near a tenth, and the executing phase is nearly half of total project time. Like the monitoring and controlling phase, activities in the closing phase occur in other phases of the project.

   c. Ideally, the closing phase, like the initiating phase, is near a tenth, the planning and controlling and monitoring phases are near a quarter, and the executing phase is near a third of total project time. Like the monitoring and controlling phase, activities in the closing phase occur in other phases of the project.

   d. Ideally, the closing phase, like the initiating phase, is near a fifth, the planning and controlling and monitoring phases are near a tenth, and the executing phase is nearly half of total project time. The closing phase does not occur until the executing phase is complete.

2. Which of the following actions fails to live up to the project manager's code of ethics?

   a. To reach a deadline, you take a shortcut and rearrange the planned order of project activities

   b. During the closing phase and after formal acceptance, you notice a bug in a software-oriented project. After internal deliberation, you contact the client, inform them of the bug, and request additional funds to fix it

   c. You realize that a team-member was not sick during a taken sick day. Since this affects the team-member's integrity and not the project manager's, you take no further action

   d. All of the above adhere to the project manager's code of ethics

3. You have recently completed the final delivery in an aerospace equipment project. The client is happy with the engine modification that was produced. As part of the closing phase, which of the following will document the successful delivery?

    a. Procurement review
    b. Product verification
    c. Procurement audit
    d. Formal acceptance

4. Which of the following is not a common cause of difficulties in an extinction-ending to a project?

    a. Resources at the end of a project taper off
    b. Once important team-members do not work on close project activities
    c. Team-members slow down work to maintain professional relationships and/or avoid being reassigned to another team
    d. There is not as much motivation to complete the close project process as compared to earlier project work

5. What is the difference between project closure and project completion?

    a. Project completion occurs at an exact point, whereas project closure is a more continuous process.
    b. Project completion refers to finishing project work, while project closure refers to evaluating all project work.
    c. Project closure happens at an exact point, whereas project completion is a more continuous process.
    d. Project closure refers to finishing project work, while project completion refers to evaluating all project work.

6. What is the correct order in closing the procurement billing process?

    a. Receive formal acceptance, close out all charge accounts, and reconcile ongoing transactions
    b. Reconcile ongoing transactions, receive formal acceptance, and close out all charge accounts
    c. Reconcile ongoing transactions, close out all charge accounts, and receive formal acceptance
    d. Receive formal acceptance, reconcile ongoing transactions, and close out all charge accounts

7. Which of the following is not a common issue that surrounds the implementation of a project closure plan?

   a. When the project closure plan is set, many team-members have completed their tasks and moved on to other assignments. Those completed activities may have loose ends that will get overlooked.
   b. A project closure plan often must be recalibrated multiple times to account for changes made to the project during the execution phase.
   c. Some project activities may be ongoing, or even just beginning, as the vast majority have been completed, meaning the project closure plan is unevenly applied.
   d. A project that is integrated or starved leads to a project closure plan being evoked far earlier than expected.

8. How could starving a project and pulling out of a contract lead to unexpectedly higher costs?

   a. If liquidated damages are applied in favor of the contractor, not the performing organization
   b. If an arbitrator decides the performing organization is more at fault
   c. If making the project a permanent part of the performing organization is costlier than planned
   d. None of the above

9. In the middle of project work, a global recession begins. The client who commissioned the project must cut the project scope drastically. Contractors declare bankruptcy and it is difficult to find replacements. The project manager's performing organization decided to combine the project with another ongoing project to consolidate costs. However, the client balked at this combination and insisted the project be completed at a reduced scope. The performing organization agreed and the project was completed in an expedited fashion.

Which of the following correctly describes the scenario above?

   a. After an integration proposal was proposed, the project was starved and completed with a reduced scope.
   b. After an addition proposal was proposed, the project was starved and completed with a reduced scope.
   c. After an addition proposal was proposed, the project was completed with a reduced scope.
   d. After an integration proposal was proposed, the project was completed with a reduced scope.

10. Which of the following is not a source of information to use in the lessons learned report for projects that end successfully?

   a. Viability decisions
   b. Progress reports
   c. Stakeholder surveys
   d. Project logs

Section 5.4: Practice exam questions on closing – answers

1. Answer: C. Initiating is 13%, planning is 24%, executing is 31%, controlling and monitoring is 25%, and closing is 7% in the standard model for project phase breakdowns. Answer C's rounding is closest to these percentages. Closing work begins in the planning phase, when the project closure plan is made. Additionally, protocol for closing should be altered in the controlling and monitoring phase if project scope is altered.

2. Answer: C. Answer A adheres to ethical project management processes. Rearranging the order of project activities is part of the course during most projects. There is no indication that established norms in the project were contravened when changing the plan. Choice B is an example of upholding product integrity. No matter how late an error is found, it should be noted. There is nothing unethical about discussing adding funds to a project. Answer C could harm upcoming project work if unaddressed. Project managers must take responsibility for the entire project and address any organizational rules that are broken.

3. Answer: D. Answers A and C deal with procurements and are unrelated. Procurement review occurs before contract acceptance and is part of the request for proposal phase, evaluating vendors. Procurement audits, meanwhile, measure the completeness of contracts and how they can be improved. Product verification and formal acceptance are similar. Product verification verifies that project work is indeed complete and matches contract requirements and/or stakeholders' expectations. The project manager is required to document formal acceptance after delivery and testing. Product verification is an internal test, while formal acceptance depends on the project sponsor's satisfaction.

4. Answer: A. Resources tapering off is associated with a project being starved to an early end, and not standard close project processes. Resources are planned out for the project close as they are for every phase. Tapering off is not the correct term. There are fewer resources in the last phase since it requires the least time. All of the other choices present common issues that are encountered in the final phase of a project.

95

5. Answer: B. Project completion is not the same as project closure. After the final delivery stage is carried out and the closure processes outlined in the last section are checked off, all project activities are complete. But between project completion and the closure date, all project work must be evaluated so lessons are learned and future projects can go smoother. Only answer B satisfies these conditions. Both completion and closure happen at an exact time when they are finished.

6. Answer: D. Project management best practices stipulate that you get formal written acceptance from the contractor before taking any steps to close out billing accounts. Of the two billing options, you cannot and should not try to close charge accounts before handling transactions with contractors or subcontractors.

7. Answer: A. Answer A implies the project closure plan should be set after project activities are completed. However, the project closure plan is set at the same time every other plan is, before project work begins. It may be altered as the project changes, but the project closure plan must begin before the execution phase begins. Every other choice is true of project closure plans.

8. Answer: B. If a contract includes a provision to have an arbitrator decide disputes, the arbitrator has the discretion to assign a costlier burden than expected upon an organization if they break a contract. The decision to end a project by starvation often involves breaking a contract. In answer A, liquidated damages are a set amount and calculated before a starvation decision is made. Answer C describes a project that ends by addition, not one that ends by starvation.

9. Answer: D. The scenario only described an integration proposal, not a starvation ending. Integration ends one project by merging it with another project or folding it into an existing work unit. Starvation ends a project by withdrawing resources until nothing remains. This project had a drastic reduction in scope after the integration proposal was rejected.

10. Answer: A. Viability decisions concern decisions on whether to continue the project and do not shed light on the course of project work. Projects that end successfully do not ever engage in viability decisions. All of the other options provide valuable insight into the rhythms of project work.

Glossary

I have partnered with Quizlet to create an app that features all the terms in the glossary. You can get the Quizlet app in the App Store or Google Play. Search for "Robert Nathan Project Management."

The url for the flashcards is: https://goo.gl/7ZRr41.

Absolutes of quality management, Absolutes of quality management is a project quality theory developed by Philip Crosby. Crosby is known for championing "zero defects" and trying to prevent redoing tasks in projects with large supply chains.

Acceptance criteria, Acceptance criteria details what constitutes project completion per project charter rules and how all parties to the contract will express their agreement that the project is complete.

Activity description, Activity description is the process by which project work is divided into activities. For larger projects, activities are tied to specific deliverables.

Activity list, The activity list details each project activity's scope.

Actual cost (AC) (or actual cost of work performed, ACWP), AC is the real amount spent for the work done in a specific date range. AC is often contrasted with planned value (PV).

Adaptive project life cycle, An adaptive project cycle is a combination of the iterative and incremental project life cycle models. Adaptive cycles must work in increments because the nature of the work is only found by doing it. The specifics of work at the end of the project are unknown at the start of the project. Adaptive project cycles depend on getting feedback after each increment of work and changing work techniques to meet that challenge.

Addition closure, Addition is a type of project closure. Instead of cleanly ending, a project becomes a permanent feature of the firm, such as evolving into separate unit within the organization with ongoing operations. Addition necessitates that more resources are added to the project to make it permanent.

Administrative closure, Administrative closure signifies that all information needed regarding the project has been received and accepted by the appropriate parties and further project communication is not expected.

Administrative costs, Administrative costs represent the cost of paying project team members' salary, servicing contract fees, and general costs that allow an organization to operate at capacity.

Agile project management, Rather than depending on unknowns or pauses, the agile project embraces flexibility. Agile projects use sprints to work through the project. Sprints are increments, short work cycles that focus on improving the project in a specific way. The work that occurs in each sprint determines the next sprint. Instead of working in the project from start to finish, agile managers work on parts in a more ad hoc manner as project needs arise. The point is to continuously improve the project and address the unexpected quickly.

Allowable payback time, The allowable payback time metric specifies the deadlines when payments relevant to project must be delivered.

Analogous cost estimating, Analogous cost estimating examines the budgets of similar completed projects within the firm or accessible budgets (for example, government contractors) from outside the firm to estimate costs.

Application area, Projects that share specific resources, business strategy, clients, or team-remembers are application areas. A project can fall into more than one application area.

Apportioned effort, Apportioned effort is project work that is understood to be continuous and not divided into discrete work packages.

Ascertained or liquidated damages, Ascertained or liquidated damages is a set amount, either fixed daily or on some other timeline, to be repaid in the event of a breached contract.

Assumption analysis, Assumption analysis finds all the implicit thinking done about the project and checks the premises for accuracy and completeness. It is important to bring the entire project team together so capabilities are not overestimated and calendars are synchronized around the project.

Arbitration, Arbitration mandates that an independent third-party come in when there is a dispute, hear all parties, and make a binding decision. It is a private legal process outside of the courtroom. An increasing amount of contracts include a recourse to arbitration.

Backward pass, The backward pass scheduling method begins at the end date to determine the longest an activity can be delayed without extending the project's entire timeline.

Balanced matrix, An organization with a balanced matrix is one in which a project manager's authority is equal to the authority of each line manager. A line manager therefore has a lot of independence in their assigned work.

Benefit-cost ratios (BCR), BCR indicates the total per-dollar value of a proposal and can be used throughout the project's life anytime an expenditure is being considered to compare options. BCR is often used when considering outside contractors.

Benefits justification, A benefits justification initiates a project because the benefits exceed the cost and effort of a project. This is informed by the NPV analysis.

Bottom-up cost estimating, Bottom-up cost estimating uses the work breakdown structure to map out a project's cost, dollar-by-dollar.

Brainstorming, Brainstorming brings together project team regulars and irregulars, including executives and third-party contractors, to map out the likeliest risks and obstacles a project will face.

Budget at completion (BAC), BAC is the total budget allocated for the project in project plans; the sum of the budget for each project phase.

Budget baseline, Budget baseline is the original, final cost estimate before project work begins.

Budget types, The three types of project budgets are: stage budget (or the money for the planned work of the stage), change budget (or the money put aside to cover changes), and risk budget (or the money put aside for potential financial impacts from risks).

Budgeted cost of work scheduled (BCWS), The BCWS provides totals for all types of specialized work over the entire scope of a project.

Burn rate, The burn rate indicates the speed with which project funds and resources are spent. All projects have a burn rate. They key is to control the burn rate so it matches budget plans.

Business case (or business justification), Every firm that initiates a project has a business case to justify the project. The business case takes the prospective project's costs, risks, and benefits into account to show that a project is worthwhile.

Causal methods (or econometric methods), Causal methods are used to identify variables that might change your project's forecast. Methods that are included in this group include econometrics, regression analysis, and autoregressive moving average (ARMA).

Cause and effect (or a **fishbone** or **Ishikawa**) **diagram,** A cause and effect diagram's purpose is to find the origins of project problems. An example of a cause and effect diagram is shown below:

Ceteris paribus, Ceteris paribus is Latin for "other things equal." It is a common technique to make decisions between two similar options.

Change control board (CCB), Usually used in larger organizations, a CCB has the responsibility to approve requested changes and set the cost and timeline for an additional change.

Code of accounts, A code of accounts is a customized numbering or lettering system that allows for a shortening of complex jargon or product identifiers.

Compliance justification, A compliance justification initiates a project because a partner organization or a cooperation's headquarters needs it done.

Configuration management, Configuration management controls the rollout of approved changes to the activities and goals of the project. The importance of configuration management is tied to stakeholders' goals for the projects. The more goals that exist, the more important configuration management becomes.

Confirmation bias, Confirmation bias is the tendency to interpret results to support a pre-existing viewpoint rather than being open to being wrong.

Conflict management, One of a project manager's greatest responsibilities is to smooth over disagreements, expedite communication, and deescalate tense situations. There are five strategies for conflict management within project management: withdraw, smoothing, direct force, reconciliation, and problem solving.

Contingency planning, Contingency planning recommends changes to make in response to a risk that occurs. Contingency plans take the form of: "if this, do this."

Contingency reserves, Contingency reserves provide slack for work most likely to go wrong and allow for minor setbacks to be easily addressed and funded.

Control charts, Control charts are mostly drawn up for industrial or manufacturing projects in which minor physical defects could set the project back. It displays the quality or grade of a good against upper and lower control limits.

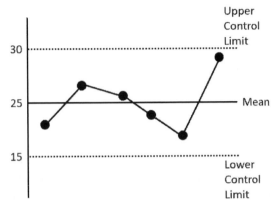

Controlling communications plan, The controlling communications plan is the subsidiary plan that structures how a project manager will conduct themselves through each type of communication: lateral, upward, and downward.

Controls review, The control review analyzes the strengths and weaknesses of how project work was regulated, whether planned controls kicked in as expected, and if they handled project changes smoothly. It also proposes ways to improve project controls. It is often part of the lessons learned report.

Core competencies, Core competencies differentiate one organization from another. Such specializations determine eligibility for a project. A firm's employees' education and experience determine core competencies. A new project will either strengthen a business's existing core competencies or create new ones.

Corrective action, A corrective action is taken whenever there is a problem in the execution of a project plan and project work is changed in order to correct the problem.

Cost management plan, A cost management plan details how the project's expenses will be determined, monitored, and controlled. It is a subsidiary project plan.

Cost of quality (COQ), COQ refers to the expenditure required to achieve a specific standard in the project.

Cost performance index (CPI), In earned value management, CPI calculates the quotient of earned value (EV) to actual cost (AC). The CPI indicates the project's cost-efficiency: how much project work is completed for the expenditure. The CPI can help prevent unexpected over-spending. If CPI is calculated to be above one, the project is under-budget. If CPI is calculated to be below one, the project is over-budget. If SPI is equal to one, the project's planned budget is being followed to the penny. The equation for CPI is:

$$CPI = \frac{EV}{AC}$$

Cost reimbursable contracts, Cost reimbursable contracts are meant for projects with high levels of risk and many obstacles. There are defined and funded costs, usually towards the start of the project, but there is a shared understanding that the costs in the latter phase of the project are uncertain. All cost reimbursable contracts see the seller receive periodic credits from the buyer for all charges.

Cost variance (CV), In earned value management, CV calculates the difference between the earned value (EV) and actual cost (AC). Allowing a project manager to quickly see the amount they are under or over budget, CV clearly shows the magnitude by which cost plans have or have not gone awry:

$$CV = EV - AC$$

Crashing, Crashing shortens the entire project's timeline by consolidating activities and adding more resources. Usually activities that take the most time are focused on, and their end-dates are brought up. Crashing adds float relative to the original timeline.

Critical chain method, The critical chain method changes the project schedule to account for scarce resources. Due to its focus on resources, critical chain method by necessity uses feeding buffers to ensure a resource can be spread around within the project team and be available when needed.

Critical path, The critical path is the sequence of deliverables that takes the most time to finish the project by using forward pass methodology. Since everything on the critical path is required, the critical path is the shortest amount of time in which a project can be completed.

Dashboard Reporting, Dashboard reporting allows a project's leaders and stakeholders to jointly review project documentation and reports. It pulls together all project reports into one, easy-to-access place. Dashboard reporting is often a part of project information systems.

Decomposition, Decomposition is the technique by which high level project requirements are broken down into lower level project work.

Define Measure Analyze Design Verify (DMADV), Stemming from six sigma standards of quality, the DMADV theory stresses the importance of data collection and checking for accuracy during each phase of project work.

Define Measure Analyze Improve Control (DMAIC), Stemming from six sigma standards of quality, the DMADC theory stresses the importance of understanding project stakeholders' needs and setting team goals around that common purpose. The theory stipulates a repeating cycle of defining, measuring, analyzing, improving, and controlling these shared expectations during project work.

Deliverable, A deliverable is what project work produces that is meant for those whom the project was undertaken. A typical project has many deliverables that are given to clients and stakeholders throughout the life of the project.

Delphi technique, Delphi technique uses outside experts independently to get a fair assessment of issues in the project or the project plan. Specialists are contacted and asked to develop an analysis of a project's strengths and weaknesses when it comes to likely risks. Experts do not know who the other analysts are or if there are other analysts at all. The results from each analyst are then compared and similarities in each report are acted upon.

Direct force, In conflict management, direct force doles out a decision to address underlying causes for team friction and demands compliance. This is not the best strategy but is often resorted to after repeated offenses and when deadlines are approaching and a more nuanced approach cannot be taken. Usually the underlying cause is not addressed, and the conflict pops up again.

Discount rate, The discount rate is used to estimate the future value of today's currency, per changes in inflation and currency exchanges. It is most relevant in international projects.

Discrete effort, Discrete effort is easily measured project work that produces a specific output.

Discretionary dependencies (or **soft logic**), Discretionary dependencies refer to a sequence of tasks not dictated by the nature of the work, but rather agreed to within the project team.

Downward communication, Downward communication occurs within the project team and concerns day-to-day tasks. New assignments, schedule updates, and reviewing others' work are the most common topics in downward communication.

Earned value (EV) or **budgeted cost of work performed (BCWP),** EV is what the plan allocated for the actual work completed. EV determines whether the project is over- or under-budget by calculating what should have been spent in a specific date range. The equation for EV is:

$$EV = BCWP = (BAC)(percent\ of\ project\ actually\ completed)$$

Earned value management (EVM or **earned value analysis**), EVM allows for a project manager to remain current with budgetary and spending plans in detail. EVM reviews a project's cost status relative to schedule status based on expenditure.

Efficient communication, Efficient communication stresses the timeliness in addition to the relevancy of communication.

Enabling justification, An enabling justification initiates a project because the project will have spillover benefits that will improve other operations.

Ending a contract by default, Ending a contract by default is due to time expiring and the product or service to be rendered is no longer needed by the buyer.

Ending a contract for cause, Ending a contract for cause is when one party breaks the contract's rules and is unable to hold the terms.

Ending a contract for convenience, Ending a contract for convenience is when one party to the contract withdraws from the contract by choice.

Enterprise environmental factors (EEFs), EFFs can be internal or external. Internal factors include a firm's culture, history of successfully managing projects, business partners, assets, and knowhow such as subject-matter expertise. External factors include location, distance, market changes, and government regulatory conditions.

Escalation of issues, If a project manager's conflict resolution efforts are not successful, then it may morph into escalation of issues in which higher levels of management are involved. Escalation of issues occurs any time during the execution of a project in which a project manager cannot get things back on track.

Estimate at completion (EAC), EAC is a forecast of the total cost of the project. EAC can be calculated at different points of the project with the current sums of actual cost (AC) and estimate to complete (ETC):

$$EAC = AC + ETC$$

Estimate at completion (EAC) forecast for work performed at present CPI, This variant of EAC is tied to calculations of work performance productivity and has a simple equation that is a result of substitutions to the EAC and CPI equations:

$$EAC_{CPI} = \frac{BAC}{CPI}$$

Estimate to complete (ETC), The ETC is a dynamic number that is calculated when a project is ongoing. An ETC should decrease as the project's end nears unless there are drastic budget changes that lead to high costs as the project matures.

Expected monetary value (EMV), The EVM expresses the monetary impact of an opportunity or risk. EMV is often used to assess projects during the initiation phase. It can also be used to evaluate decisions with a project. The EMV is an amount that is either a profit (a positive number) or a loss (a negative number):

$$EMV = (profit\ probability)(expected\ profit\ amount)$$
$$- (loss\ probability)(expected\ loss)$$

Expected value (EV), EV is used by organizations to figure out the positives or negatives associated with a risk. Expected value is the numerical product of probability a risk materializes and the impact of the risk. Ranking risks' expected value from those close to zero to those close to one is known as ordinal scale. Impact is the amount of harm or help a risk carries. Impact is expressed on a scale from 0 (very little effect on the project) to 1 (critical effect on the project). Probability is the percent chance the event occurs. The equation for EV is:

$$expected\ value = probability\ x\ impact$$

Exploit strategy, An exploit strategy is executed when there is a positive effect to a risk occurring, such as monetary savings, and acting in such a way to realize that positive outcome.

Extinction, Extinction is the ideal type of project closure. Extinction implies a project was successfully completed and accepted. An extinct project does not linger on and has a definitive end point.

External dependency, An external dependency is an event that occurs outside the project scope that triggers a schedule change.

External failure costs, External failure costs occur when the customer, project sponsors, or regulatory agency outside the firm find the product or process to be inferior. External failure costs operate as external risks before they occur.

Fast-tracking, Fast-tracking requires that two tasks originally scheduled to be done in an order to be instead performed at the same time.

Feasibility study, Looking at the required abilities to fulfill a project, a firm should decide whether a project is one they can or cannot manage. If it is determined feasible, the team in charge of the feasibility study should estimate the time, resources (new and existing), and individuals needed to complete the project.

Feeding buffers, Feeding buffers allow for certain activities that feed into the critical path more time than needed to be completed, as to ensure they do not delay many deliverables depending on them.

Fifty-fifty technique, Used in earned value management, the fifty-fifty technique is a mix of the percent-complete and milestone techniques. In fifty-fifty, a project activity is zero percent before relevant work begins, fifty percent while work is ongoing, and one hundred percent complete once work is finished.

Finish to finish (FF), FF describes a precedence diagram relationship where one activity cannot finish until another one finishes.

Finish to start (FS), FS describes a precedence diagram method relationship where one activity cannot start until another one finishes.

First unit, A first unit is the first complete product the project produces before producing many more of them.

Fitness for use, Fitness for use is a project quality theory developed by Joseph Juran. Juran is most concerned with design, safety, and conformance. He developed a ten-step program to implement what he dubbed "quality by design."

Fixed price contracts, Fixed price contracts are mostly employed for detailed projects with few risks or where there is a high level of trust between the concerned parties. Fixed price contracts are contrasted with cost reimbursable contracts.

Float (or slack), Float refers to the latest possible delay a deliverable can be started without delaying the project.

Force majeure exemption, The force majeure exemption allows a contract's obligations to be nullified when outside conditions prevent the contract's work from being undertaken (an "act of god").

Forward pass, The forward pass scheduling method calculates early start and end dates, moving forward from the starting node going through each deliverable.

Forward projections, Forward projections use past project data, such as productivity and how far from reality project plans proved to be, to predict the future course of the project and the effects of all the changes to project scope.

Forward view, A forward view is a control process by which a project manager lets the project team know what metrics they will be checking to measure progress the next time the project is reviewed.

Free float, Free float refers to the amount of time a activity can be postponed without delaying the start of another, separate activity.

Fully burdened rate, The fully burdened rate is the total amount needed to sustain all of a project team's resources, from utility bills to every salary.

Functional organizational structure, A functional organization is divided into specialized departments, where each department works in one area instead of cooperating towards a larger goal.

Gantt chart, A Gantt chart has a vertical axis representing time and horizontal bars reflecting how long a project task is projected to take. An example of a Gantt chart is below:

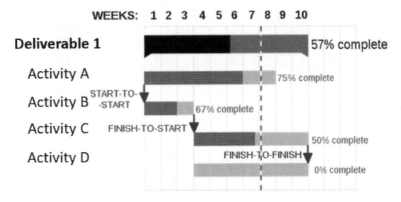

Graphic evaluation and review technique (GERT) chart, GERT charts use network analysis for complex projects, often in engineering, and allow feedback loops, repetitive steps, alternative paths, and probabilistic and conditional branching. The numbers in a GERT are attained by running mathematical Monte Carlo simulations. These simulations sample outcomes using random methods to try to quantify the totality of possible outcomes. An example of a GERT chart is below:

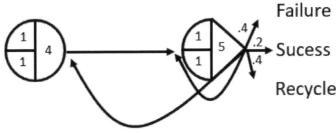

Grade quality, Grade quality differentiates quality not by cost, but by technical requirements and capabilities.

Handover protocol, Handover protocol is a project closing process that ensures all necessary project products and artifacts have been successfully signed off on and documents archived correctly.

Hierarchy of needs, Hierarchy of needs refers to Abraham Maslow's motivation theory. It stipulates that people need to feel they are actualizing their potential to be truly invested in an endeavor.

Histogram, A histogram is a bar graph in which the horizontal axis is an occurrence category and the vertical axis is that category's frequency. An example of a histogram is below:

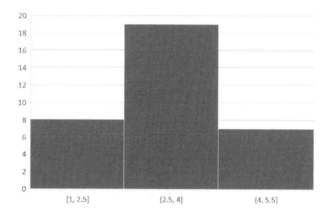

Human resource management plan, The human resource management plan defines roles and assignment-owners within the team. It sets the hierarchical order within a project to establish lines of communication and reporting requirements.

Incremental project life cycle, An incremental project life cycle is one in which feedback is prioritized. Work pauses frequently in order to gather stakeholder opinion. During these pauses, project work is tested and feedback is incorporated. There is a cycle of increments of project work and pauses to evaluate that increment before the next increment of work can begin.

Indemnity, An indemnity protects parties to a contract against future loss by requiring a reimbursement in the event of project cancellation.

Integration, Integration ends one project by merging it with another project or folding it into an existing work unit. Projects that end in this way see their resources (such as equipment) and team-members either assigned to different projects or returned to their prior roles before the project began.

Integration management, Integration management is concerned with coordinating the competing needs of stakeholders and ensuring the tradeoffs of making a certain choice are understood.

Integrated project management software, Integrated project management software provides a central hub to list and assign tasks to tagged team members, provide comments and feedback, allow work progress monitoring and feedback, and usually some integration with word processing, databases, and other programs via attachments or inbuilt programs.

Intermediate steps, Intermediate steps are not major milestones but day-to-day activities. Keeping track of intermediate steps helps a project manager track progress through the life of the project and keep the project on schedule.

Internal failure costs, Internal failure costs occur when a product or process is still in development, but is deemed by the project team to be inferior and in need of being reworked.

Internal dependency, An internal dependency is an event that occurs within the project team, but is not tied to the execution of tasks, that can change the project schedule.

Interpersonal factors, Interpersonal factors refer to the project team's experiences, skills, team dynamics, and cultural sensitivities. All of these factors need to be improved to ensure project work is smooth.

Issue log (or **change log**), The issue log handles all extra work and attaches owners and due dates to each new project task. Communications dealing with large project activities should also be documented in an issue log.

Iterative project life cycle, An iterative project life cycle is one in which the timelines and costs change as project work continues. The project scope is known, but the sequence of the work cannot be planned out. An iterative approach is often used when developing a product for the first time and there are unknowns in the production process.

Knowledge areas, Project management's ten knowledge areas are management strategies that shape project activities from the first to last day: communication management, cost management, human resources management, integration management, procurement management, quality management, risk management, scope management, stakeholder management, and time management.

Key performance indicators (KPIs), KPIS are measurements of project progress and effectiveness. There are dozens of options for KPIs and a project manager must decide which KPIs to include in the project. Common KPIs include earned value, Gantt charts, and performance management plans.

Lag, A lag extends the time between two dependent tasks. A lag is necessary when two preceding tasks feed into one successor task and one of these preceding tasks is completed much earlier than the other.

Lateral communication (or **cross-organizational communication**), Lateral communication is conducted through peers, such as a project manager, contractor, and functional manager in another department. Usually, everyone in a lateral communication is heavily invested in the project.

Latest finish date, The latest finish date uses the backward pass to find the furthest off date in which the project work can be completed while still in line with the relationship and dependencies between project activities.

Lead, A lead modifies a relationship between project activities and accelerates work on a dependent successor task when the preceding task is bogged down. A lead is a scheduling technique.

Level of effort (LOE), LOE is used to determine how long a project activity will take and depends on the skillset of the team-member. LOE is measured in time and not necessarily tied to a specific output.

Line or **functional managers,** Line managers oversee a set of deliverables that requires specialized skillsets. Line managers assist project managers by allowing them to make informed decisions about resource allocation, scheduling, or costs. Line managers often use subject matter experts.

Lessons learned report, A lessons learned report uses the project reports to review the project, with a focus on corrections made during project work. Beyond narrating the process by which errors were made and plans altered and the results of those corrections, the report includes valuable insights learned during the project generally. Also known as a post-project evaluation, post-project reviews, or project postmortem.

Linear responsibility chart (LRC), A LRC is a type of responsibility assignment matrix (RAM) which ties deliverables or activities to individuals on the project team to provide clear lines of responsibilities.

Lines of communication, Lines of communication refer to the total number of possible communications between each member of a project team.

Maintenance justification, A maintenance justification initiates a project to provide an update or improvement to a past project.

Major process groups, Project management's major process groups are initiation, planning, executing, monitoring and controlling, and closing.

Make or buy analysis, Make or buy analysis is a critical step in the procurement process. Many large firms have the resources to "make" many goods or services in-house while smaller firms have to "buy" more outside help. But the decision to make or buy is a decision about whether to initiate the procurement process.

Management reserves, Management reserves are for unforeseen obstacles and are therefore not funded in the cost baseline. Management reserves are funded from a different account and used if the project scope changes or expected resource costs changes.

Managing by exception, Managing by exception focuses on small breaks with project plans to prevent larger issues from occurring.

Mandatory dependencies (or **hard logic),** In scheduling, tasks in a mandatory dependency have to be done in an exact order.

Master schedule, A master schedule sets due dates for all deliverables and highlights milestones around which deadlines cluster.

Matrix, Businesses with a matrix organization use many resources for multiple reasons at the same time and there is competition for resources and unclear lines of communication.

Mature project management, Mature project management relies on a corporate culture, established methodology, and deep understanding of project-managing fundamentals to see through successful projects. The more completed projects and lessons learned reports an organization has, the more mature its project management expertise should be.

Maximax criterion (or Hurwicz criterion), Maximax criterion emphasizes the possibilities inherent in a risk—it is the "half-full" view.

Maximin criterion (or **Wald criterion),** Maximin criterion emphasizes the losses possible in risks—it is the "half-empty" view.

Milestone or **project phase,** A milestone is a crucial development in a project. Completing a large deliverable or realizing a deliverable that marks the end of the project are examples.

Milestone technique, In earned value management, the milestone technique measures project progress by the total amount of projective activities that are complete.

Mitigation planning, Mitigation planning offers strategies to implement in order to make a risk less likely from occurring or to contain the risk's effects if things start to worry the project manager.

Motivation-hygiene theory, Motivation-hygiene theory is a work performance theory developed by Frederick Herzberg. Herzberg believed that if people expect to be rewarded after completing an assignment they would do it better and motivation is external.

Multi-criteria decision analysis, Multi-criteria decision analysis helps determine the likelihood of a new team-member jelling with your existing staff.

Needs theory, David McClelland's needs theory hones in on achievement, power, and affiliation as three things team-members need to believe they possess to dedicate themselves to a project. McClelland's model posits team motivation is internal.

Negative risk, A negative risk is a threat that will possibly impede project progress, such as losing out on a key resource during a deadline.

Net present value (NPV), The NPV compares the value of investing in one project over another by finding the profit difference. If the NPV yields a positive amount, the project should result in a net gain in profitability or value to the company. If NPV is a negative amount, the project manager has determined the project will amount to a loss and should not recommend initiation.

Nodes, Nodes are individuals, departments, or some other entity that communicate with the project manager during the project.

Non-project-based firms, Non-project-based firms, like manufacturers who produce similar products over time, are not organized for project after project. Such businesses require an unfamiliar and upfront effort to put in place rules to complete a project.

Normal distribution, In a normal distribution, one standard deviation includes everything 34% above and below the mean (68% total), while two standard deviations includes everything within 47.2% of the same mean (95% of total).

Opportunity cost, Opportunity cost is the loss incurred when one path is taken over another. It is the best path not taken; the alternative, optimal use of the time that you used on a task.

Organizational factors, Organizational factors refer to the performing organization's culture, hierarchy, risk management, financial restrictions, safety standards, and rules governing how departments interact with one another. All of these factors need to be used to improve project work.

Organizational project management, Organizational project management focuses on how programs and portfolios of many projects should be regulated to achieve a larger strategic goal.

Organizational risks, Organizational risks stem from in-fighting and competition for resources within an organization that often is juggling multiple projects. The larger organization rejects the project's scope and budget once the project begins and needs resources.

Overhead costs, Overhead costs are those things that are not tied to specific project work but instead cover basic operations. Team members' health insurance, office supplies, renting office space, and utility costs are examples of overhead.

Parametric cost estimating, Parametric cost estimating uses mathematical formulas to estimate costs of a deliverable by using the relationships between variables like per unit cost.

Pareto diagram, Pareto diagrams have two distinct vertical axes. On the left vertical axis are frequency measurements representing how often the phenomena on the horizontal axis appeared. On the right vertical axis are cumulative percent measurements of each occurrence from left to right. Below is an example of a Pareto diagram:

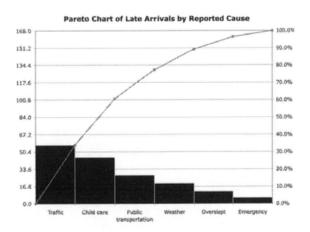

Performance appraisals, Performance appraisals encompass the many forms reviews of project work can take, including formal and informal methods.

Percent-complete technique, In earned value management, the percent-complete technique estimates the exact amount of the project completed. This method requires the most meticulous project monitoring to not lead to an erroneous estimate.

Performance measurement plan, The performance measurement plan enables the tracking of work done in achieving goals and the evaluation of those on the project team.

Performance risks, Performance risks stem from an organization depending on untested methods, people, corporate partners, or technology to fulfill project requirements. Performance risks can also be labeled as technical or quality risks.

Plan-do-check-act (PDCA) cycle, A theory from W. Edwards Deming and Walter Shewhart, the PDCA cycle targets continuous improvement of processes and production. The key is the PDCA cycle's four steps to be done repeatedly: every action (the last part of the cycle) should lead to another plan to improve that action.

Plan-do-study-act (PDSA) cycle, The PDSA cycle is a project quality theory developed by W. Edwards Deming. Deming thought of quality as chiefly the responsibility of project managers. The PDSA tests a change and widely shares the results so a method can be tinkered with and improved. The PDSA is action-oriented and involves experimentation.

Planned value (PV) (or **budgeted cost of work scheduled, BCWS),** PV is the planned budget for work in a specific date range.

Portfolio, A portfolio groups many projects together for a common strategic, long-term reason.

Positive risk, A positive risk is an opportunity to possibly achieve an objective that will help the firm, like completing a project with less resources.

Performing organization, The performing organization is the organization whose staff is most directly involved in planning and executing the project. The project manager works for the performing organization. The client may create a contract with the performing organization to regulate the project.

Precedence diagram method (PDM), The PDM maps out a schedule's interrelationships. The PDM uses boxed nodes that represent activities and arrows to represent the sequence and dependencies between each activity. There are four primary PDM relationships: finish to start, finish to finish, start to start, and start to finish.

Privy mutual relationship, The privy mutual relationship of a contract specifies that the requirements of a contract only apply to the parties specified in a signed contract, not to outside organizations.

Problem solving, In conflict management, problem solving requires the project manager to introduce a new set of guidelines to permanently address underlying issues. The belief is there is one correct solution and it must be found.

Problem updates, Problem updates provide ongoing information in identified project shortcomings.

Process improvement plan, A process improvement plan hones in on inefficiencies, can help mitigate the chance of incurring failure costs, and can help a project meet cost estimates.

Procurement file, The procurement file is the collection of all records dealing with contracts attached to a project.

Product risk, Product risk occurs when a new product, including software and hardware products, fails to live up to expectations. This failure may be due to scheduling issues with unexpected technical slowdowns, or resource issues due to either compatibility or capability shortcomings in equipment and facilities.

Procurement closure, Procurement closure is the process by which a contract is terminated, either successfully or unsuccessfully. Procurement closure can be done many times during a project, depending on how many contracts were associated with the project.

Procurement statement of work (P. SOW), The P. SOW is the plan for procurement process. It identifies the purchases the project needs as it progresses and how team-members can request additional purchases. Procurement plans should define which resources will be received internally from within the organization, which resources will require an outside partner, and the expectations for delivery of relevant resources.

Product verification, Product verification is an ongoing process that is only completed when the project is closed. Product verification verifies that project work is indeed complete and matches contract requirements and/or stakeholders' expectations.

Productivity updates, Productivity updates specify the rate of task-completion per resource input. Productivity information can allow a project manager to project progress forward and extend or collapse deadlines.

Programs, Programs are groups of related projects that are initiated simultaneously to achieve efficiencies. These projects require similar work. Grouping them together allows for a firm to specialize.

Program evaluation review technique (PERT) charts, PERT Charts display the relationships between project activities. Nodes are deliverables in projects and arrows represent events, as in network diagrams. Arrows are not drawn at scale to represent time, which in the example here is measured in months. To determine the time between project tasks, PERT employs probability to estimate the total duration of the project. Below is an example of a PERT chart:

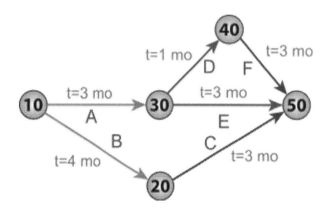

Program evaluation and review technique (PERT), A PERT tries to determine accurate timelines before project work begins. In its simplest form, a PERT weighted equation is used to find the mean estimate using optimistic (O), pessimistic (O), and most likely timelines (M from mean) for the entire project or a specific deliverable:

$$PERT = \frac{O + 4M + P}{6}$$

Project, A project is an undertaking that has a distinct start and end date. A project's goal is to produce a new good or service. Completing a project will require a combination of inputs: natural resources, labor, capital, and/or entrepreneurship. A project's schedule, product, and resources are its main elements.

Project activity, A project activity is a specific portion of the project work that has separate parts that are similar. Different project activities often require different skillsets and personnel. Activities can include deliverables and work packages.

Project life-cycle costing, Project life-cycle costing compiles all costs every project task will accumulate over its lifespan. It groups categorical costs, such as acquisitions, operations, and disposals, and compares alternative work flows to find savings.

Preassignment, Preassignment skips or at least expedites the process of requesting and reviewing outside firms in the procurement process. Often, a buyer will have a partner organization in mind and offer them the entire contract or part of it first.

Project-based or **projectized firms,** Project-based firms, like consultancies, usually have multiple ongoing projects at a time and an existing strategy for tackling projects and a project manager has more authority than in a non-project based firm. In projectized organizations, teams that work on a project are often ad hoc and lack a functional base that keeps them together after a project is complete.

Project champion, A project champion is someone higher than the project manager who strongly advocates for the project. A project champion can cajole team-members into reordering their priorities to expedite a project's completion and fight for more resources.

Project charter, The project charter offers an understanding of the project's core goal and offers a preview of project participants and objectives. Once the project charter is signed, the project may begin. The project charter gives the project manager the authority to begin work.

Project checklist, A project checklist is an ongoing list of remaining project activities at various levels of the project. The checklist can contain basic work packages or large deliverables. Checklists usually do not mix higher and lower level tasks within the same list.

Project closure plans, Project closure plans are those that set the budget and timelines for activities in the last phase of a project. They are set in the planning phase like every other project plan and need to be recalibrated to project changes.

Project controls, Project controls represent the processes by which project progress is regulated and accepted.

Project coordinator, A project coordinator assists the project manager who is responsible for the overall project. The project manager can delegate to the coordinator responsibility for various tasks. In large projects, a coordinator is responsible for keeping everyone on the same page.

Project expeditor, A project expeditor's main responsibility is communication. A project expeditor works under the project manager and cannot make any decisions on their own authority.

Project forecasting, Project forecasting uses past project information to predict future performance, conditions, and task completion timelines.

Project information systems (PMIS), PMIS is an online database created for the shared use of the project team. The PMIS enables project managers to organize and disseminate progress reports on each facet of a project. There are specialty PMIS for specific types of projects. Building information modeling (BIM) is a common PMIS used in the construction industry, for example.

Project initiator, The initiator is the person that requested the project for a specific reason. The project sponsor may or may not be the same person. The project initiator, whether inside or outside a project manager's firm, should be integrated into early discussions on communication protocols.

Project life cycle, The project life cycle is the sum of each project activity, from start to finish. The five stages of a project life cycle are: initiation, planning, executing, controlling and monitoring, and closing.

Project management code of ethics, The code of ethics focuses on four areas that allow a project manager to do the right and most effective thing: responsibility, respect, fairness, and honesty. The code takes precedence over everything else.

Project manager, A project manager is responsible for the day-to-day planning, execution, monitoring, and closing of a project. A project manager reports on progress and timetables to project stakeholders, contractors, and higher management within the performing organization.

Project management office, In larger firms with many ongoing projects, a project management office handles staffing on multiple different projects and allows individuals working on multiple projects to prioritize their work.

Project management plan, A project management plan is the collection of many specialized and subsidiary plans that regulate project work.

Project management risks, Project management risks stem from poor scheduling, budgeting, and inconsistent methodology. These risks arise when a project plan is found to be extremely inaccurate during the execution phase.

Project monitoring, Project monitoring deals with day-to-day managing and figuring out underlying strengths and weaknesses as the project plan is executed.

Project scope, The project scope details all the work that must be done to complete a project. The scope is determined in negotiations between the project client and the performing organization. The scope is essential for cost-estimating. Project problems usually arise when the scope is enlarged or diminished.

Project sponsor, A project sponsor is above the project manager within the same organization, sets deadlines, oversees funding, and provides guidance during the project's duration. Project director, project executive, or a senior responsible owner (SRO) are other terms for the project sponsor.

Project statement of work (SOW), A SOW defines the project scope and the strategic vision behind the project.

Project steering group (PSG), When appointed, A PSG is responsible for approving project work in one phase before allowing progress to the next stage. Important stakeholders, higher management, and project sponsors make up a PSG.

Preliminary project scope statements, Preliminary project scope statements describe a project's objectives, the desired deliverables, and anticipated timelines before a firm commits to a project. All necessary deliverables required to complete the project are termed critical success factors in the preliminary project scope.

Quality control plan, A quality control plan specifies what acceptable and unacceptable standards for project work are.

RACI chart, RACI is an acronym for: responsible, accountable, consulted, and informed. It designates who needs to be responsible, accountable, consulted, and/or informed at specific points of project work. It is a type of responsibility assignment matrix (RAM).

Rebaselining, Rebaselining changes the scope of a project. Rebaselining occurs when substantial changes to project plans occurs or whenever a new project plan is executed or critical path project activities are altered.

Reconciliation, In conflict management, reconciliation attempts to have aggrieved parties compromise on something, share the blame, and agree on a solution.

Regression analysis cost estimating, Regression analysis cost estimating looks for statistical trends between the costs of similar deliverables to estimate costs at varying scales.

Remote teams, Remote teams are separated and communicate virtually. Team members in this case are still committed to the project, and many work on a daily basis. Remote teams are contrasted with teams in a tight matrix organization.

Reserves, Reserves are resources or assets that are meant to only be used in the event of a risk occurring. Reserves are sometimes built into a project's budget to assist in risk-management. Reserves can take monetary and non-monetary forms.

Reserve analysis, Reserve analysis is the monitoring process used to assess resource reserves and decide whether and which additional resources are needed, or whether some resources are inefficient and not needed.

Resource breakdown structure (RBS), The RBS divides resources up into categories and is an essential tool in cost estimates. Resources common across many projects include: salaries, rent, subject matter expert fees, and online utility costs.

Resource calendars, Resource calendars are developed during the planning stage and guide the execution phase's development. Resource calendars display team-members' assignments and when they are scheduled to work on project activities, as well as when non-human resources will be available.

Resource crashing, Resource crashing adds additional resources to expedite a project's critical path.

Resource leveling (or **resource-based methodology**), Resource leveling occurs when multiple activities need the same resource at the same time, or a resource is available only at certain times, meaning that a project manager must look to ensure resources are allocated by priority within the project schedule.

Resource smoothing, Resource smoothing changes activities' float times to increase resource availability.

Responsibility assignment matrix (RAM), A RAM ties deliverables or activities to individuals on the project team to provide clear lines of responsibilities.

Requirements traceability matrix, A requirements traceability matrix links project activities to ensure that project work can be validated as complete by the correct protocol.

Reverse resource allocation, Anytime a resource is only used once and is dependent on other activities, reverse resource allocation should be used to schedule the resource's use. A project manager works in reverse because task dependencies can be more easily checked from the end date.

Reworks, Reworks are improvements to subpar products and project outcomes generally and are another type of scope change.

Risk appetite, Risk appetite details the level of unpredictability project leaders can accept based on resources. Project decisions and gambles are determined by how great the risk appetite is and the corresponding budgetary flexibility.

Risk audits, Risk audits are continuous checks on the project's progress from start to finish to identify upcoming obstacles or ongoing underperformance. Auditors should have access to the project manager to keep tabs on the project.

Risk planning processes, Risk planning processes create and put procedures in place to deal with late or inefficient work or anything else that negatively impacts the project.

Risk register, The risk register is a list of risks ordered by priority and including each risk's probability, likely cost, and timeline. Risk registers are diagramed, often as scatterplots, and try to ensure the budget accounts for project risks.

Risk thresholds, Risk thresholds define the range within which project leaders are willing to take a chance. Certain risks could result in unacceptable costs. Some risks pose costs potentially so great, they are beyond risk thresholds and therefore will never be attempted.

Risk tolerance, Risk tolerance details the level of unpredictability project leaders can accept based on risk-reward calculations. Tolerance will be higher if project leaders reason that the reward possibilities outweigh the potential costs.

Risk tracking, Risk tracking encompasses the methods a project manager will employ to record risk activities. Developments are logged before the threat materializes, and tracking will encompass action to minimize costs.

Risk transfer, Risk transfer occurs when the risk has not occurred and the potential consequences are taken up by a third-party. Purchasing insurance for various possibilities is a form of risk transfer.

Rolling wave planning, Rolling wave planning details project plans more over time—the project is done in waves. With a rolling wave approach, the project begins purposely with many unknowns that are only defined by subsequent project work.

Root cause analysis, Root cause analysis tries to look past symptoms of repeated failure and find perhaps unnoticed issues. Team chemistry, communication protocol, or organizational weaknesses are all fair game in this approach.

Rule of seven, Rule of seven is a rule of thumb in sampling. The rule states that seven examples are needed to make a real trend and ensure you are not falling for anecdotal or misrepresentative evidence.

Schedule management plan, The schedule management plan pulls team and budget planning together and defines the flow that will structure the project.

Schedule network analysis, Schedule network analysis is a scheduling technique to analyze the duration and dependencies of project activities. It often produces a network diagram to display the relationships between activities.

Schedule performance index (SPI), In earned value management, SPI calculates the quotient of earned value (EV) to planned value (PV). SP is an easy way to see if the project is ahead of schedule or behind schedule and how efficient project work to a certain point has been. The equation is:

$$SPI = \frac{EV}{PV}$$

Schedule variance (SV), In earned value management, SV calculates the difference between a planned budget for a portion of the project and the actual amount spent for the same portion of the project. The SV can determine the extent to which you are behind, ahead, or up-to-date with the schedule by showing the gap between expected and actual completion. The equation is:

$$SV=EV-PV$$

Scope change, All authorized changes to a project plan are a scope change.

Scope creep, Scope creep comprises all unauthorized changes to the project plan that are not authorized as they happen.

Scope statement, The scope statement specifies the results the project will produce and what the firm requires to do the work in writing. The statement should specify the project team, costs, deadlines, and contingency plans for handling unexpected events.

Scrum framework, Part of iterative projects, scrum is a rugby formation in which all of the team huddles together and tries to gain momentum in the game. The scrum framework dictates that what is known about the project at the start is negligible. Instead of having a plan, the team should be tightly focused on understanding the facts of a project and advancing together through conditions that may or may not change. In the scrum framework, incremental plans are tested for next steps but there are no large-scale, start-to-finish commitments. The scrum master plays the part of the project manager in this framework.

Secondary risk , A secondary risk arises in response to decisions made about a prior risk, such trying to exploit it.

Share strategy, A share strategy is a form of risk transfer that deals with comparative advantage and turning risks into opportunities. It involves outsourcing risks to a third-party. Rather than the motivation being insuring and preventing future loss, the partnership in a share strategy is to increase overall revenue.

Six sigma (6σ) measurements, 6σ measurements is a project quality theory developed by Bill Smith and Mikel J. Harry. Smith and Harry focus on output and believe an optimum number of allowable defects rounds out to 3.4 per 1,000,000. 6σ quality also informs the Define Measure Analyze Improve Control and Define Measure Analyze Design Verify theories.

SMART Objectives, Specific, measurable, aggressive, realistic, and timeline-driven objectives, will help keep the team cohesive and moving forward.

Smoothing, In conflict management, smoothing tries to create breathing space in the short-term to see if an issue blows over. It minimizes problems and tries to reinforce ground-rules in communication that may have lapsed.

Stage-gate process, The stage-gate process is used when there are risks about doing many activities at the same time. "Gates" are set up at strategic points to monitor work. Before work can continue, the activity's progress must satisfy a checklist before the project can pass through the gateway to the next set of activities.

Stakeholder, A stakeholder is anyone whose interest in the project must be taken into account during project work.

Stakeholder grids, Stakeholder grids classify the power and influence of each stakeholder, from least to most in whatever category is being measured. The four grids are: the influence/impact grid, the power/influence grid, the power/interest grid, and the salience model. Salience refers to the power, legitimacy, and urgency of each stakeholder.

Stakeholder registers, Stakeholder registers identify, categorize, and evaluate stakeholders based on expertise and/or interest in specific areas of the project. The three steps of stakeholder analysis are: 1) identify stakeholders; 2) analyze potential impact; 3) assess stakeholders' likely reactions.

Start to finish (SF), SF describes a precedence diagram relationship where one activity cannot finish until another one starts.

Start to start (SS), SS describes a precedence diagram method relationship where one activity cannot start until another one starts.

Starvation, Starvation ends a project by withdrawing resources until nothing remains. A decision to starve a project implies failure and an unplanned early end. Since even in the most drastic situations work must taper off in some form, even quickly ended projects are considered to be starved.

Sub-critical activities, Sub-critical activities are project activities that can be skipped in a critical path.

Summary activity or **hammock,** A summary activity looks to find related activities that can be grouped together to streamline later project work.

Systems management, Systems management is a framework for understanding how part of a project affects the whole.

Theory X managers, Theory X managers are hands-on. They believe people avoid working hard and take short-cuts. Theory X managers stress discipline and rules and subscribe to the motivation-hygiene theory, doubting their workers want to succeed because they like the work.

Theory Y managers, Theory Y managers are hands-off. They believe people work up to expectations and outperform when held accountable. Theory Y managers subscribe to the needs theory and do not micromanage.

Theory Z managers, Theory Z management was developed by William Ouchi and stresses setting. Ouchi saw motivation stemming from environment and job security. If team-members feel confident that their work is stable and they are free to take the occasional risk, their morale would be higher and performance improved.

Thomas-Kilmann Conflict Mode Instrument (TKI), The TKI offers another way to achieve team cohesiveness. The TKI uses a questionnaire to determine how team members act in tense situations. With this data in hand, project managers can adapt conflict management techniques to their team to resolve issues.

Three Is of project management, The three Is recognize that every project task is integrated, interconnected, and interdependent with all others. The three rule underlines the complexity of project work.

Three-point cost estimating, Three-point cost estimating is the average of the likely, worst-case, and best-case cost estimates.

Tight matrix, Teams in a tight matrix are those that are co-located and work with each other daily in the same physical space. A tight matrix can take the form of a weekly meeting where everyone is physically present, or daily interactions of some sort in a shared space.

Time phased budget, A time phased budget is a budget in which funds are only released when prior project activities are completed. Time-phased release is a common condition in a project contract.

To-complete performance index (TCPI), In earned value management, TCPI determines the cost performance needed to meet the project's planned budgets, or the budget at completion (BAC). TCPI looks at remaining project tasks and comes up with the amount these tasks must come in under at to keep the project within the budget's confines. The equation is:

$$TCPI = \frac{BAC - EV}{BAC - AC}$$

Top-down cost estimating, Top-down cost estimating looks at only the top levels of the WBS and estimates by milestone.

Total float, Total float refers to the amount of time an activity can be postponed and begun later without pushing back the overall project end date.

Total quality management (TQM), TQM is a project quality theory developed by Armand Feigenbaum. Taking a systematic view of an organization that is cohesive, TQM encourages a constant mindset to continuous improvement to provide increasingly higher quality products.

Trade-off analysis, Trade-off analysis aims to evaluate the effects of moving resources away from one activity to another in order to optimize the upside of a trade-off. Most trade-offs are competing ones: changing the resource distribution between competing activities invariably helps one and hurts the other.

Triggers (or risk symptoms), Triggers are indications that a risk will materialize into a threat.

Triple constraint, The triple constraint is between the competing basic areas of a project: scope/product, deadlines/schedule, and resources/inputs. A change in one of those three areas will have effects on the other two.

Upward communication, Upward communication consists of a project manager reporting on project progress to executives or influential stakeholders.

Validate scope process, The validate scope process inspects project documentation and output for accuracy and completion and validates each deliverable as complete.

Value engineering, Value engineering optimizes project performance and cost by primarily looking to eliminate unnecessary costs.

Variable sampling, Variable sampling measures and compares quality on a scale to measure the level of difference from product to product.

Variance at completion (VAC), At the end of a project, VAC calculates the difference between the planned budget (the estimate at completion or EAC) and the actual budget at completion (BAC):

$$VAC = BAC - EAC$$

Viability decision, A viability decision is made about whether to continue or end the project early in response to a project failure. A project sponsor and the larger organization make the decision, and it is up to the project manager to organize and execute the starvation closing plan.

Waterfall management, Waterfall management is the most popular type of project management. In contrast to iterative forms, the waterfall methodology follows discrete phases to complete a project. Like the course of a waterfall, the project proceeds through tried and true phases. Sequentially, it goes through the process groups: initiation, planning, executing, controlling and monitoring, and closing.

WIIFM ("what's in it for me"), WIIFM ("what's in it for me") is a phrase project managers should think about when dealing with project stakeholders. Stakeholders should be given clear reasons for caring about project progress.

Withdraw, In conflict management, withdraw implies a conflict is ignored rather than addressed. This could occur when one party refuses to discuss the conflict and merely adjusts, a party leaves the project, or it is decided the conflict was a minor issue that does not warrant attention.

Workarounds, Workarounds are responses to unexpected events that are not planned but required to solve a pressing issue or negative risk event. Often, it is up to a project manager to decide on the specific workaround. By definition, workarounds are not included in the risk register since they are unforeseen.

Work breakdown structure (WBS), A WBS defines the scope of the project, focuses on deliverables, shows how deliverables are related to one another, and is the basis for planning after a project is initiated. The WBS decomposes the work needed to satisfy deliverable requirements.

Work package, A work package is the most detailed day-to-day description of project work. Project duration and cost are determined by analyzing each work package.

Each of the following two hundred questions has one answer. You have four hours to complete the exam.

1. Which of the following is an example of qualitative risk analysis?

    a. Calculating the damages various risks could inflict
    b. Gathering all of the possible risk events a project faces
    c. Prioritizing risks by potential damage
    d. Planning on ways to avoid risks

2. The largest portion of a project's budget is typically spent in which phase?

    a. The execution phase
    b. The closing phase
    c. The monitoring and controlling phase
    d. The planning phase

3. Within an organization that is managing a portfolio of projects that share resources, multiple projects are running behind. A project manager for one of these projects is growing frustrated with her project being ignored and resources taken away. Which of the following could help relieve this situation?

    a. Resource levelling
    b. Updating team communication protocols
    c. Appeals to higher management
    d. Contingency planning

4. During the monitoring and controlling phase, what actions should a project manager take to address the issue of a team member not responding to emails?

    a. Plan on either decreasing or increasing the team member's future pay based on performance
    b. Clearly state the team member's duties ahead of time
    c. Ensure the team member's functional manager understands the scope of the project
    d. None of the above

5. Which of the following requires a project manager to seek a legal opinion?

    a. A stakeholder is discovered to have a conflict of interest with work done for one project deliverable
    b. An outside contractor requests access to proprietary information from the project manager's firm which had hitherto not been shared
    c. A project team member claims emotional trauma from overtime work
    d. All of the above

6. Which of the following statements is true?

    a. Programs are groups of related deliverables
    b. Portfolios group projects together for a common strategic, long-term reason
    c. Hammocks group together overlapping project phases
    d. Project life cycles consist of overlapping project milestones

7. The risk register, human resource management plan, scope baseline, and cost management plan are all required in which project planning activity?

    a. Developing the schedule
    b. Defining activities
    c. Estimating costs
    d. Estimating activity duration

8. A project manager has identified twelve risks for a deliverable and categorized them in a risk register, involved the relevant stakeholders, developed ways to track the risks, and rated them by priority. What would be the logical next step in the project manager's analysis?

    a. Negotiate higher risk thresholds
    b. Smooth out conflicts between parties that share risk
    c. Identify the triggers for each risk
    d. Determine if the risk is insurable or not

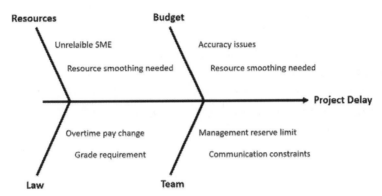

9. The above diagram's purpose is to:

    a. Find the origins of project problems
    b. Reflect how long a project task is projected to take
    c. Display the relationships between project activities
    d. Provide network analysis for complex projects

10. Your team member is two days late with a report. Just before a meeting in which the report's topic will be the main subject discussed, he emails it to you. You immediately notice that the report has serious errors. What should you do?

    a. Hold the meeting and allow the team member to present the report, errors and all
    b. Hold the meeting and inform your team members there are errors in the report
    c. Delay the meeting until the report has been corrected by the team member
    d. Delay the meeting and correct the report yourself

11. A project statement of work (SOW) includes which of the following?

    a. The summary activity, a list of all deliverables, and strategic vision behind the project
    b. A list of all deliverables and the project scope
    c. A list of all deliverables, list of all team members, and the project scope
    d. The project scope and the strategic vision behind the project

12. Who has access to the stakeholder register?

    a. Only the project manager
    b. The listed stakeholders
    c. All team members and stakeholders
    d. Anyone with whom the project manager shares the register

13. What are the benefits of a RACI chart?

    a. It lowers risk by making responsibilities clear
    b. It shows which tasks team members spend time on, even if it is a small amount
    c. It shows the roles of the project team in various project activities
    d. All of the above

14. A code of accounts assists which aspect of project managing?

    a. Time management because a code of accounts simplifies jargon in communication
    b. Hammocking because a code of accounts is ordered in a way to logically sequence tasks
    c. Time management because a code of accounts simplifies cost estimates
    d. Decomposition because a code of accounts allows an easy way to breakdown deliverables

15. How do building information modeling (BIM) and project information systems (PMIS) relate?

   a. Both refer to online collaborative environments meant to increase team cohesiveness and are launched during the execution phase of a project
   b. Configuration management is a part of BIM and PMIS is focused on dynamic scheduling
   c. Both refer to online collaborative environments meant to increase team cohesiveness and are launched during the initiation phase of a project
   d. BIM is focused on dynamic scheduling and PMIS is focused on stakeholder engagement

16. You are 40% of the way done with a project to upgrade a bank's internal network. There are 100 locations in Canada and another 50 in the United States. An accounting software vendor has just released a significant software upgrade for most of the equipment you are installing. This software upgrade would offer the customer increased functionality that they initially said they needed that did not exist when the project began. What should you tell the bank?

   a. Let the bank know about the upgrade and its potential impact on the project's timeline and functionality if implemented
   b. Install the upgrade and change the timeline as necessary, since the bank said they needed this type of functionality to begin with
   c. Install the upgrade on the remaining sites and keep your timeline on schedule
   d. Tell them nothing and continue along the same trajectory. The bank did not anticipate the update

17. You have recently been promoted to project manager because the last project with a specific client resulted in dissatisfaction and higher costs than expected. This new project requires you to create a system that ensures a manufacture's products satisfy consumer safety laws. The manufacturer expects this system to quickly reduce his cost. Which of the following is not true in this situation?

   a. Assumptions about consumer safety laws should be evaluated to ensure the system effectively meets standards
   b. The objective is to do better than the last project manager
   c. The schedule is a constraint partly because the owner expects a quick result
   d. The budget is a constraint partly because the last project came in over budget

18. A project manager leads a team of twenty people. Some individuals communicate daily, and others are silent for days at a time. More often than not, the project manager is carbon-copied on emails. How many lines of communication exist in this project?

    a. 10
    b. 20
    c. 105
    d. 210

19. Which of the following statements about change control is true?

    a. The change control team should first record and categorize client requests for a change
    b. Fixed-price contracts make change controls less necessary
    c. Regression plans should be created during the assessment phase of change control
    d. Change control is the same thing as configuration management

20. What is the difference between regression analysis cost estimating and parametric cost estimating?

    a. Regression analysis cost estimating looks for statistical trends between the costs of similar deliverables to estimate costs at varying scales. Parametric cost estimating uses mathematical formulas to estimate costs of a deliverable by using the relationships between variables like per unit cost.
    b. Parametric cost estimating looks for statistical trends between the costs of similar deliverables to estimate costs at varying scales. Regression analysis cost estimating uses mathematical formulas to estimate costs of a deliverable by using the relationships between variables like per unit cost.
    c. Regression analysis cost estimating is the average of the likely, worst-case, and best-case cost estimates. Parametric cost estimating examines the budgets of similar completed projects within the firms or accessible budgets (for example, government contractors) from outside the firm to estimate costs.
    d. Parametric cost estimating is the average of the likely, worst-case, and best-case cost estimates. Regression analysis cost estimating examines the budgets of similar completed projects within the firms or accessible budgets (for example, government contractors) from outside the firm to estimate costs.

21. Which of the following is not a characteristic of adequate project scoping?

    a. A requirements traceability matrix is developed to allow understanding of how deadlines were decided

    b. Project assumptions are listed and tested as the project develops

    c. The project is broken down into work package level to anticipate workloads

    d. Measurement criteria for many project requirements are determined

22. During the controlling of project work, a project manager submits a report on project progress to other contractors to review. As the report circulates for further review, the project manager notices that data he entered earlier had been altered at multiple places. The data is altered to overstate project progress and outside contractors were not involved in the altered areas. What action should the project manager take?

    a. Reenter the data to be accurate and only take further action if it is altered again

    b. Email the contractor most likely to have altered the data to see if they did it

    c. Email every contractor to discover who altered the data

    d. Notify higher management to examine the issue

23. Which of the following statements is not true?
    a. Programs are groups of related projects managed together

    b. Portfolios are groups related deliverables

    c. Hammocks group together overlapping project tasks

    d. Project life cycles consist of overlapping project phases

24. Both simple average and PERT calculations can help determine:

    a. Three-point estimation

    b. Project scheduling

    c. Sequencing activities

    d. Estimating activity durations

25. During the planning phase, a functional manager cleared ten team-members for project work three months down the line. When the time arrives, the functional manager only provides five workers and says her team will need double the scheduled time. Which of the following outlines the appropriate response?

    a. If the relevant work is on a critical path, evoke contingency plans and resource reserves to find ways to prevent work delays
    b. If the relevant work is on a critical path, appeal to higher management to get more resources
    c. If the relevant work is not on a critical path, evoke contingency plans and resource reserves to find ways to prevent work delays
    d. If the relevant work is not on a critical path, appeal to higher management to get more resources

26. In which of the following organizational forms does project management hold less sway than expertise?

    a. Matrix
    b. Laissez faire
    c. Type X
    d. Functional

27. What is the chief difference between the program evaluation review technique (PERT) and the critical path method (CPM) charts?

    a. CPM prioritizes dependencies between tasks while PERT prioritizes duration between tasks
    b. PERT prioritizes dependencies between tasks while CPM prioritizes duration between tasks
    c. CPM uses most likely duration while PERT uses mean of estimates
    d. PERT uses expected value while CPM uses most likely duration

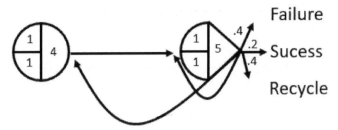

28. The numbers in this chart come from:

    a. Probability multiplied by impact
    b. Running mathematical Monte Carlo simulations
    c. Determining the number of nodes of communication
    d. Subtracting the probable expected loss from the probable expected profit

29. In which of the following contracts is the profit unknown?

    a. Cost plus contract
    b. Cost reimbursable contract
    c. Fixed price contract
    d. Time and materials contract

30. A nanotechnology firm needs to use one of the world's fastest supercomputers during one phase of an upcoming project. The supercomputer charges by the hour. The project manager works backwards from the project's project completion date to best schedule the time when the supercomputer will be used. The project manager is employing which of the following resource-scheduling techniques?

    a. Reverse resource allocation
    b. Resource crashing
    c. Resource leveling
    d. Resource smoothing

31. What is the difference between the percent-complete technique and the milestone technique?

    a. The percent-complete technique calculates that project activity is zero percent before relevant work begins, fifty percent while work is ongoing, and one hundred percent complete once work is finished, while the milestone technique measures project progress by the total amount of projective activities that are complete, with partially complete activities not counting towards the percent of project actually completed calculation.

    b. The percent-complete technique measures project progress by the total amount of projective activities that are complete, with partially complete activities not counting towards the percent of project actually completed calculation, while the milestone technique estimates the exact amount of the project completed. This method requires the most meticulous project monitoring to not lead to an erroneous EV estimate.

    c. The percent-complete technique measures project progress by the total amount of projective activities that are complete, with partially complete activities not counting towards the percent of project actually completed calculation., while the milestone technique calculates that project activity is zero percent before relevant work begins, fifty percent while work is ongoing, and one hundred percent complete once work is finished.

    d. The percent-complete technique estimates the exact amount of the project completed. This method requires the most meticulous project monitoring to not lead to an erroneous EV estimate, while the milestone technique measures project progress by the total amount of projective activities that are complete, with partially complete activities not counting towards the percent of project actually completed calculation.

32. A project manager has just finished reviewing the mandatory relationships between project activities. The project manager was engaged in which of the following processes?

    a. Developing the schedule
    b. Defining activities
    c. Sequencing activities
    d. Estimating activity durations

33. Which of the following qualities is the least important quality of an effective project manager?

    a. Unfaltering self-reliance
    b. Changing plans after failures occur
    c. Productive and efficient communication
    d. Facilitating interaction

34. How does the procurement plan relate to other project plans?

    a. It makes up a large part of the budget and cost management plan
    b. It shapes the human resource plan by defining the level of expertise needed to work with various resources
    c. It can determine the schedule's sequencing of tasks by revealing where there are resource constraints
    d. All of the above

35. Clark is a project manager that oversees a team of eight that has worked together on more than a dozen projects. For their latest project, his team has determined that there are 235 risks and 12 major causes of those risks. The project work is nearly complete and has been approved by all key stakeholders, and the client has been very supportive and involved. Despite all this, there is one risk the team cannot determine an effective way to prevent or insure against, and the work cannot be outsourced or removed. What should Clark do?

    a. Determine a way to transfer the risk
    b. Determine a way to avoid the risk
    c. Determine a way to mitigate the risk
    d. Accept the risk

36. All projects share which of the following characteristic(s)?

    a. A start point and end point
    b. Cyclically repeating activities
    c. A project coordinator on staff
    d. All of the above

37. Within a profit-maximizing firm, which of the following situations uses the opportunity cost concept best?

    a. During the budget-planning process, a project is estimated to total out at $300,000 and net a profit of $50,000. The project is taken because other possible projects yield a net less profit

    b. During the budget-planning process, a project is rejected because the project will yield a negative profit

    c. During the budget-planning process, the project manager compares costs of analogous projects to see which is the best opportunity

    d. During the budget-planning process, a firm considers all the ways to cut costs while maintaining quality

38. In engineering or IT projects, "configuration management" is a synonym for:

    a. Quality control
    b. Version control
    c. Quality assurance
    d. Project assurance

39. A "post-mortem" is used for a project's:

    a. Closure
    b. End
    c. Evaluation
    d. Completion

40. You are taking over from a project manager that moved to another department early on in the initiation phase. The departing project manager tells you that she was working on share strategies and triggers. The project manager was working on which of the following?

    a. The cost management plan
    b. The quality management plan
    c. The risk management plan
    d. The procurement management plan

41. Jorge is a project manager who also happens to be a technical expert in his project's field, information technology. His project revolves around installing new software onto the computers of five separate departments of a Fortune 500 company. He also has a background in communications and management. The project has been going relatively well, and the software is resulting in many positive and necessary benefits for the company. However, a significant and unexpectedly high number of changes to project plans have been made as client feedback is analyzed. Which of the following reasons is the most likely cause of this issue?

    a. The project needs additional management oversight since has so many benefits to the company
    b. The project management team needs to use more project management communication processes
    c. Some stakeholders were not identified by Jorge and his team
    d. Jorge and his team do not fully understand the company's environment

42. The chairperson of the project steering group is the:

    a. Team leader
    b. Project manager
    c. Project user
    d. Project sponsor

43. The four types of cost-reimbursable contracts are:

    a. CPFF, CPIF, CPAF, CPEC
    b. CPFF, CPIF, CPOF, CPPC
    c. CPFF, CPIF, CPAF, CPPC
    d. CPEF, CPIF, CPAF, CPEC

44. A project manager is overseeing project activities done consecutively and is nervous about the risks of planning later project activities before knowing more details. It is decided that the project plans will be finalized piece by piece, as project work sheds light on future requirements. Planning out project work in this manner is known as which of the following?

    a. Rolling wave planning
    b. Stage-gate process
    c. Decomposition
    d. Expert judgment

45. Which of the following is NOT a part of the scope baseline?

    a. Project scope statement
    b. WBS
    c. Project charter
    d. Scope statement

46. Which of the following is not a grid that can be used to classify the power and influence of each stakeholder?

    a. Power/impact grid
    b. Influence/impact grid
    c. Power/influence grid
    d. Salience model

47. What are the three main types of project budgets?

    a. Stage budget, risk budget, change budget
    b. Stage budget, change budget, project budget
    c. Maintenance budget, risk budget, change budget
    d. Stage budget, change budget, maintenance budget

48. Which of the following is NOT one of the three costs associated with cost of quality (COQ)?

    a. Prevention
    b. Appraisal
    c. Completion
    d. Failure

49. In the wake of an economic recession, the US government created the program Car Allowance Rebate System (CARS), more affectionately known as "Cash for Clunkers," in 2009. This program provided Americans with monetary incentives to trade in their less fuel-efficient vehicles for new, more fuel-efficient vehicles. If you were a project manager working for an automotive company back in 2009, which of the following methods would analyze the way CARS would change a project's forecast?

    a. Time Series Methods
    b. Judgmental Methods
    c. Econometric Methods
    d. None of the above

50. Which of the following could be included in a work performance report?

    a. Time completion forecast
    b. Risk and issue status
    c. Results of variance analysis
    d. All of the above

51. Penelope has just started as a project manager for a large project that is half-way into its one year run. The project has 22 members on her project team, and also involves 10 sellers. In order to quickly appraise how the project is currently doing, which type of report should Penelope review?

    a. Progress
    b. Work status
    c. Forecast
    d. Executive

52. What is the difference between calculating cost variance and cost performance index?

    a. Cost variance calculates the difference between the earned value and actual cost while cost performance index calculates the quotient of earned value to actual cost
    b. Cost variance calculates the difference between the earned value and actual cost while cost performance index calculates the quotient of earned value to planned value
    c. Cost variance calculates the quotient of earned value to planned value while cost performance index calculates the difference between the earned value and actual cost
    d. Cost variance calculates the difference between the earned value and actual cost while cost performance index calculates the quotient of earned value to planned value

53. Anna is a first time project manager creating a procurement statement of work. A seller will be doing most of the actual work. Some stakeholders think that the procurement statement of work should only include the functional requirements while others want as many items as possible added. What would you advise Anna do in this situation?

    a. Determine if the seller has greater expertise than the buyer, and if so, only include functional requirements
    b. Determine if the seller has greater expertise than the buyer, and if so, be as detailed as possible
    c. Ensure the procurement statement of work is as detailed as needed for the project itself
    d. Keep the procurement statement of work as general as possible to allow for later clarification

54. Which of the following is not a benefit of a RACI chart?

    a. It sets accountability for project activities
    b. It clearly demonstrates decision-making authorities
    c. It projects the roles of the project team in various project activities
    d. It illustrates the sequence and dependencies between various project activities

55. Configuration Status Accounting refers to:

    a. Accounting for the status of funds for deliverables
    b. Accounting for the status of changes to the specifications of deliverables
    c. Accounting for the status of redirected of funds for deliverables
    d. None of the above

56. Which of the following is NOT one of the three steps of stakeholder analysis?

    a. Identify stakeholders
    b. Analyze potential impact
    c. Assess stakeholders' likely reactions
    d. Consider stakeholders' biases

57. When should the project evaluation report be completed?

    a.   Final delivery stage
    b.   Closure stage
    c.   Project completion
    d.   Project closure

58. The _____ is the collection of all records dealing with contracts attached to the project.

    a.   Final review
    b.   Procurement file
    c.   Project postmortem
    d.   Controls review

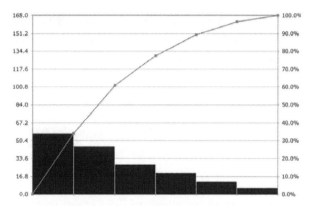

59. For a Pareto diagram like the one pictured above, the numbers on the left vertical axis represent ____, while on the right vertical axis the numbers represent ____.

    a.   On the left vertical axis are cumulative measurements of each occurrence from left to right. On the right vertical axis are frequency percent measurements representing how often the phenomena on the horizontal axis appeared.
    b.   On the left vertical axis are frequency measurements representing how often the phenomena on the horizontal axis appeared. On the right vertical axis are cumulative percent measurements of each occurrence from left to right.
    c.   On the left vertical axis is the total amount needed to sustain all of a project team's resources. On the right vertical axis are cumulative percent measurements of how many funds have been spent from left to right.
    d.   On the left vertical axis are frequency measurements representing the allotted amount of time in which a project can be completed in hours. On the right vertical axis are cumulative percent measurements of a project's completion.

60. Many organizations create this type of board to authorize change requests:

    a. Technical assessment board
    b. Change control board
    c. Configuration control board
    d. All of the above

61. How is a configuration management plan different from a change management plan?

    a. Configuration management plans deal with changes to parts of a product of the project, while change management plans deal with the project process
    b. Configuration management plans deal with the project process, while change management plans deal with changes to parts of a product of the project
    c. Configuration management plans deal with how you define, monitor, control and change the project schedule while change management plans deal with how you define, monitor, control and change human resources
    d. Configuration management plans deal with how you define, monitor, control and change human resources while change management plans deal with how you define, monitor, control and change the project schedule

62. Stakeholders are most critical in which of the following project activities?

    a. Budget plans
    b. Writing a lessons learned report
    c. Procurement negotiations
    d. Conducting a feasibility study

63. Which of the following is a correct sequence of project areas to focus on, from earliest to latest?

    a. Project schedule, communication requirements, risk identification, and project budget
    b. Communication requirements, project budget, project schedule, and risk identification
    c. Project budget, project schedule, risk identification, and communication requirements
    d. Project schedule, project budget, communication requirements, and risk identification

64. When does the initiation phase of a project end?

   a. When the project charter is signed by all parties
   b. When a project is being considered
   c. When project feasibility is validated
   d. As project monitoring begins

65. A formerly productive team-member has fallen behind. The quality of her work has also declined. Communication does not seem to be getting through to her. What would a theory z project manager prioritize to improve this situation?

   a. The consequences of not meeting project standards, such as getting fired
   b. The quality of the team member's past work and how valuable she is to the team
   c. Get higher management involved to prioritize the situation
   d. Improve the team-member's work environment

66. Project work that clarifies future work and enables project management to be more exacting is:

   a. Controlled
   b. Monitored
   c. Iterative
   d. Management by exception

67. Project work is about half way finished and the project charter has been altered multiple times. It looks like the work to meet the next set of deliverables will be beyond the agreed-upon scope and require more charter alternations. Who has the ultimate responsibility to sign off on changes to the charter?

   a. The steering committees
   b. The project manager
   c. The project client
   d. The project manager's supervisors

68. Which of the following illustrates what W. Edwards Deming contributed to cost of quality (COQ) understanding?

   a. Total quality management
   b. Plan-do-check-act cycle
   c. Six sigma
   d. Continuous improvement

69. Which of the following might be a way of conducting a procurement performance review?

    a. Examining the quality of contract work
    b. Quality audit
    c. Inspection of documents
    d. All of the above

70. Any authorized change to a project plan is a scope change. On the other hand, _____ comprises all unauthorized changes to the project plan that are not authorized as they happen.

    a. Scope reworks
    b. Scope issues
    c. Scope intervention
    d. Scope creep

71. When controlling for quality, taking a fraction of the total to draw conclusions about the total is called:

    a. Forward projections
    b. First unit
    c. Rebaselining
    d. Sampling

72. You are part of a project that has five teams and serve as team 3's leader. Team 2 has missed several critical deadlines in the past. This has repeatedly caused team 3 to need to crash the critical path. To help resolve this issue, you should meet with:

    a. The project manager
    b. The project manager and management
    c. Team 2's leader
    d. The project manager and team 2's leader

73. What is the ideal type of project closure?

    a. Extinction
    b. Addition
    c. Integration
    d. Starvation

74. Project managers can check that project work is indeed complete after a project ends by conducting a _____.

    a. Product verification
    b. Project verification
    c. Procurement verification
    d. Final verification

75. In Frederick Herzberg's motivation-hygiene theory, hygiene factors deal with:

    a. The satisfaction garnered from doing the work and mastering the many skills required in doing that work
    b. The environment broadly understood and includes things that prevent dissatisfaction, such as pay, social ties with team-members, and a pleasant work setting
    c. That people need to feel they are actualizing their potential to be truly invested in an endeavor
    d. Achievement, power, and affiliation as three things team-members need to believe they possess to dedicate themselves to a project

76. Of the following conflict management methods, which is generally the most time consuming?

    a. Problem solving
    b. Direct force
    c. Withdraw
    d. Reconciliation

77.

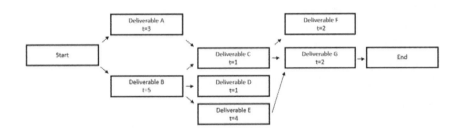

The above chart is known as a:

    a. Schedule network analysis
    b. Histogram
    c. Control chart
    d. Network diagram

78. The weather, market, and government regulatory conditions that can affect the firm are all considered:

    a. Organizational factors
    b. External EEFs
    c. Interpersonal factors
    d. Internal EEFs

79. A consultancy is most likely a:

    a. Non-project based firm
    b. Matrix organization
    c. Projectized firm
    d. Horizontally integrated

80. _____ is the technique by which high level project requirements are broken down into distinct project activities.

    a. Decay
    b. Dissolution
    c. Decomposition
    d. Disintegration

81. Which of the following is an example in which a work authorization system might NOT need to be used?

    a. The project is a very complex
    b. The project requires a lot of players
    c. The project is small
    d. The project requires approval from multiple levels

82. All of these are ways that rolling wave planning shapes work breakdown structure except:

    a. Milestones and targets are further detailed as the project evolves over time
    b. The first WBS has subproject placeholders instead of specific timelines for project completion
    c. It allows teams to work on different phases at the same time
    d. Preventing scope creep

83. The difference between earned value (EV) and actual cost (AC) is the formula for:

   a. Schedule variance
   b. Cost variance
   c. Variance at completion
   d. None of the above

84. Who is responsible for monitoring a procurement based contract?

   a. Buyers
   b. Sellers
   c. Buyers and Sellers
   d. Lawyers

85. Which of the following correctly defines risk transfer and risk thresholds?

   a. A risk transfer occurs when there is a positive effect to a risk occurring, such as monetary savings, and acting in such a way to realize that positive outcome. A risk threshold arises in response to decisions made about a prior risk, such as an exploit strategy.
   b. Risk transfer occurs when the risk has not occurred and the potential consequences are taken up by a third-party. Risk thresholds define the range within which project leaders are willing to take a chance.
   c. A risk transfer occurs when there is an opportunity to move the risk outside of the organization. A risk threshold arises in response to prior project experiences.
   d. Risk transfer occurs when a different project team is put into a riskier project area. Risk thresholds define the range within which project leaders are willing to take a chance.

86. A _____ is a program in which projects are grouped together for a common strategic, long-term reason, rather than just efficiency.

   a. Portfolio
   b. Hammock
   c. Milestone
   d. Summary activity

87. Acquiring the project team is a part of which project management process group?

    a. Planning
    b. Executing
    c. Monitoring and controlling
    d. Closing

88. Which of the following is not a technique used in conflict management?
    a. Smoothing
    b. Direct force
    c. Discrete effort
    d. Reconciliation

89. _____ planning recommends changes to make in response to a risk that occurs. _____ planning offers strategies to implement in order to make a risk less likely from occurring or to contain the risk's effects if things start to worry the project manager.

    a. Contingency, Mitigation
    b. Mitigation, Tolerance
    c. Contingency, Tolerance
    d. Mitigation, Contingency

90. Which of the following is not a need that can result in a project?

    a. Strategic opportunity
    b. Market demand
    c. Lower interest rates
    d. Customer request

91. What is the difference between a project and an operation?

    a. A project is temporary and has a defined beginning and end date, while operations are ongoing and produce the same end result repeatedly
    b. An operation is temporary and has a defined beginning and end date, while projects are ongoing and produce the same end result repeatedly
    c. Projects and operations are interchangeable terms for a series of actions with a defined beginning and end date that produce a product, result or service
    d. Projects and operations are interchangeable terms for a series of ongoing actions that produce a product, result or service

92. Iterative life cycles are a good choice for this type of project:

   a. Large projects
   b. Small projects
   c. Simple projects
   d. Projects where deliverables do not need to be delivered incrementally

93. A configuration project management plan is most important in which scenario?

   a. An ongoing manufacturing process is being repeatedly altered to be in line with multiple managers' efficiency standards. The managers do not try to reconcile their frequently incompatible ideas.
   b. A large construction project is rolling out new communication protocols. The protocol involves multiple organizations and contractors.
   c. As project monitoring begins for a newly developed software, stakeholders need to be contacted. Many of the stakeholders have not kept up with project progress.
   d. During project execution, there are many disagreements about resource sharing. Team-members complain the project schedule did not take their other work into account.

94. During testing for a newly developed educational curriculum for a school district, a school administrator asks the project manager to swap the order of two sections underneath one of thirty subheadings in the curriculum. The project manager could easily implement this change. What should the project manager do?

   a. Examine the effects of this change on other project areas
   b. Go through the standardized process of implementing any change
   c. Bring the principal's request up to the relevant team members as soon as possible
   d. All of the above

95. Which of the following is not a project activity during the project initiation phase?

   a. Designating a project manager
   b. Setting communication requirements
   c. Conducting a feasibility study
   d. Producing a scope statement

96. Which of the following is not a category of project selection methods?

    a. Mathematical models
    b. Decision models
    c. Benefit measurement methods
    d. Calculation models

97. Which of the following is not a requirements gathering technique?

    a. Brainstorming
    b. Planning sessions
    c. Focus sessions
    d. Focus groups

98. Two techniques used to determine alternative work structure or improve the project in general are:

    a. Life cycle costing and value engineering
    b. Life cycle engineering and value costing
    c. Control costing and estimate engineering
    d. Estimate costing and control engineering

99. What does the term "historical information" refer to in project management?

    a. Archived information on the current project on previous stages of the project
    b. Archived information on a closed project that is archived for reference for future projects
    c. Historical background on aspects such as human capital, funding, or suppliers, for example, that can be used to help inform the current project
    d. None of the above

100. During which project process group is stakeholder influence the lowest?

    a. Planning
    b. Executing
    c. Monitoring and controlling
    d. Closing

101. During which project process group should costs be their highest?

   a. Planning
   b. Executing
   c. Monitoring and controlling
   d. Closing

102. At a certain point in project work concerning the total project and one deliverable, the following is true:

   - The budget at completion is $2000
   - The planned value is $400
   - The actual cost is $390
   - The earned value is $500

Based on that information, which of the following is correct?

   a. The cost variance is -$110
   b. The cost performance index is 1.25
   c. The schedule variance is -$1600
   d. The schedule performance index is 1.25

103. Which of the following is NOT an input to the define activities process?

   a. Scope baseline
   b. Rolling wave planning
   c. Schedule management plan
   d. Organizational process assets

104. When can you start the planning stage?

   a. Once the outline has been approved
   b. When the project steering group or its equivalent decides planning can start
   c. Once a planning workshop has been conducted
   d. Once the stage plan has been approved

105. Which of the following descriptions describes an enabling project justification?

    a.   The project has benefits that exceed the cost and effort of the project
    b.   The project must be run because headquarters requests that it is
    c.   The project will allow other operations to produce benefits
    d.   The project must be done as a type of maintenance

106. An example of a project reserve is:

    a.   Project staff that enter a project only when key project staff are called away for other work
    b.   Activities that can be reduced in duration if the project is under time pressure
    c.   Backup materials used if needed equipment is not available
    d.   Additional sites for the project or parts of the project if the current site is unavailable

107. A project memo should be between a project manager and:

    a.   The steering group
    b.   Team leaders
    c.   Team members
    d.   All of the above

108. What is an authority trigger point?

    a.   A part of a risk plan in which it is determined when stakeholders must meet to discuss a new or evolving risk
    b.   A part of a risk plan in which it is determined when a project manager must inform the project steering group of a new or evolving risk
    c.   A part of the risk plan in which it is determined when a team leader must notify a project manager of a new or evolving risk
    d.   None of the above

109. In what order should activity processes be performed?

    a. Define activities, sequence activities, estimate activity resources, estimate activity durations
    b. Define activities, estimate activity resources, estimate activity durations, sequence activities
    c. Define activities, estimate activity durations, estimate activity resources, sequence activities
    d. Sequence activities, define activities, estimates activity resources, estimate activity durations

110. Your firm has completed the work on a deliverable as specified by carefully following the procurement statement of work. The product has been formally accepted. After one month however, your client is displeased with the results. Legally, your firm's contract is:

    a. Incomplete
    b. Complete
    c. Waived
    d. Null and void

111. At what point is risk the highest during a project?

    a. The start of project work
    b. Midway through project work
    c. Close to the project's completion
    d. It depends entirely on the project

112. Contested project changes can also be known as:

    a. Claims
    b. Appeals
    c. Disputes
    d. All of the above

113. _____ is a forecast of the total cost of the project.

    a. Actual total costs
    b. Estimate at completion
    c. Variance at completion
    d. Budget at completion

114. If your stakeholders are very hands-on and plan on actively participating in many aspects of the project, which life cycle category would be most appropriate to implement?

    a. Waterfall
    b. Iterative
    c. Predictive
    d. Agile

	SME 1	SME 2	SME 3
Activity 1	RA	I	C
Activity 2	C	R	AI
Activity 3	R	A	C
Activity 4	A	CI	R

115. Which of the following is not true about the chart above?

    a. The chart is a type of RAM
    b. The chart indicates who is notified when an activity is completed
    c. The chart should include the whole project management team
    d. The chart allows a project manager to analyze and control lines of communication

116. The four benefits of meeting quality requirements are lower costs, higher productivity, increased stakeholder satisfaction and:

    a. Increased customer base
    b. Less rework
    c. Quicker turnaround
    d. Fewer errors

117. Which of the following is produced when a project's activities are defined?

    a. Activity list, activity attributes, milestone list
    b. Activity list, activity attributes, milestone attributes
    c. Activity list, milestone activities, milestone list
    d. None of the above

118. When deriving activity duration estimates, you might round to the nearest
____, depending on the project.

   a. Hour
   b. Day
   c. Week
   d. All of the above

119. Which of the following is the equivalent of earned value?

   a. The budget at completion multiplied by planned percent of project
      completed
   b. The budgeted cost of work performed (BCWP)
   c. The actual cost at percent of project actually completed
   d. None of the above

120. Which stakeholder is typically the publisher of the project charter?

   a. Project manager
   b. Project client
   c. Planning group
   d. Executive manager

121. Which of the following is not a common result of risk management?

   a. Recategorizing risks if triggers accumulate
   b. Reserves built into the project's budget are released
   c. A share strategy is executed
   d. Subgroups of the project management team are identified as the risk
      management team

122. What is the difference between risk appetite and risk tolerance?

   a. Risk appetite details the level of unpredictability project leaders can accept based on risk-reward calculations, while risk tolerance details the level of unpredictability project leaders can accept based on resources
   b. Risk appetite defines the range within which project leaders are willing to take a chance, while risk tolerance details the level of unpredictability that is acceptable based on Monte Carlo numbers
   c. Risk appetite details the level of unpredictability project leaders can accept based on resources, while risk tolerance details the level of unpredictability project leaders can accept based on risk-reward calculations
   d. Risk appetite details the level of unpredictability project leaders can accept based on risk-reward calculations, while risk tolerance defines the range within which project leaders are willing to take a chance

123. Project scope summary and the strategic vision behind the project is a part of which of the following?

   a. A portfolio
   b. A work breakdown structure (WBS)
   c. Every project contract
   d. A project statement of work (SOW)

124. When is a project considered complete?

   a. When all of the deliverables are completed
   b. When project funds are exhausted
   c. When all stakeholders' needs are addressed
   d. When the project manager decides the project is completed

125. A project manager works for a catering company who is providing catering for a large conference. Although all the food for the conference is prepared and has been delivered to the conference hall, your company can't set up food stations until the project manager coordinating the conference has confirmed that all of the tables and booths are set up. This is an example of which of the following task dependencies?

   a. Finish to start (FS)
   b. Finish to finish (FF)
   c. Start to start (SS)
   d. Start to finish (SF)

126. A project manager is overseeing project in which not all deliverables were able to be mapped out at the start of the project because they are dependent on external factors. As a result, the project manager has subproject placeholders instead of specific timelines for project completion. Therefore, the project manager's work breakdown structure is updated incrementally. This is an example of which of the following planning techniques?

a. Rolling wave planning
b. Stage-gate process
c. Decomposition
d. Expert judgment

127. Projects with a net present value (NPV) less than zero should be:

a. Accepted
b. Rejected
c. Paused
d. Terminated

128. What constitutes project completion per project charter rules is deemed:

a. Acceptance protocol
b. Acceptance criteria
c. Handover protocol
d. Handover criteria

129. A project to revamp a company's onboarding processes is nearly complete. A big, final deliverable is an online tutorial portal. There are weeks of testing conducted to ensure the tutorial program meets standards and expectations before delivery occurs. After the testing is completed, the client praises the tutorial as above and beyond their expectations and brings up future opportunities for collaboration.

In the above scenario, the weeks of testing before delivery are considered:

a. Procurement audit
b. Product verification
c. Procurement review
d. Formal acceptance

130. Which of the following should NOT be directly considered when creating a human resource (HR) management plan?

    a.   Environmental factors
    b.   Organization factors
    c.   Personnel policies
    d.   Critical success factors

131. Albert is a project manager at an automobile company. He is in charge of updating a compact car for the upcoming year. He and his team are developing potential changes to the car and are listing these changes' quantifiable and non-quantifiable benefits. Of the benefits listed, which of these is a non-quantifiable benefit?

    a.   Switching air-conditioning manufacturers will save the automobile company 15% in costs
    b.   The sleeker redesign of the car's body will attract new customers
    c.   The compact car will take two hours less total to manufacture
    d.   The updated audio and screen display will allow the company to charge customers $1,500 more MSRP

132. When utilizing the milestone technique, a partially complete project activity:

    a.   Counts towards the percent of project actually completed calculation
    b.   Counts as halfway towards the percent of project actually completed calculation
    c.   Does not count towards the percent of project actually completed calculation
    d.   Counts only the exact amount of the project activity completed towards the percent of project actually completed calculation

133. Consider the formula: $\frac{BAC-EV}{BAC-AC}$. This formula used for:

    a.   To-complete performance index
    b.   EAC forecast for work performed at present CPI
    c.   One-time cost variance EAC
    d.   Repeated cost variance EAC

134. A project manager compares costs from previous projects to extrapolate costs for a current project. This is an example of:

    a. Bottom up estimating
    b. Three point estimating
    c. Parametric estimating
    d. Analogous estimating

135. Which of the following is not a form of communication that should show up in an issue log?

    a. Upward communication
    b. Lateral communication
    c. Controlling communication
    d. Downward communication

136. _____ protects parties against future loss by requiring a reimbursement in the event of cancellation.

    a. Arbitration
    b. Viability decision
    c. Indemnity
    d. Ascertained damages

137. A project manager leads a team of 18 people. How many lines of communication exist in this project?

    a. 342
    b. 171
    c. 162
    d. 324

138. Critical path method (CPM) charts are most similar to:

    a. GERT charts
    b. Gantt charts
    c. PERT charts
    d. RACI charts

139. The chairperson of a project steering group is the:

a. Team leader
b. Project manager
c. Project user
d. Project sponsor

140. _____ is project work that is understood to be continuous and not divided into discrete work packages. By contrast, _____ is easily measured project work that produces a specific output. _____ is project work measured in time and is not necessarily tied to a specific deliverable.

a. Discrete effort, apportioned effort, level of effort
b. Apportioned effort, level of effort, discrete effort
c. Level of effort, discrete effort, apportioned effort
d. Apportioned effort, discrete effort, level of effort

141. Which of the following is a shared property of project closure and project completion?

a. They are both continuous processes and do not have an exact date
b. Depending on the project, they occur at the same time
c. They both depend on the project's stakeholders
d. All of the above

142. Handoffs can also be known as:

a. Technical transfers
b. Feasibility changes
c. Phase sequences
d. None of the above

143. Your business case can be produced by considering the project's:

a. Justification
b. Benefits
c. Roles
d. Scope

144. Which of the following is a common property of a project's procurement process?

    a. Requests to perform
    b. Invitations to auction
    c. Requests for quotation
    d. Invitations to inform

145. Which of the following is not a project activity during the project planning phase?

    a. Norming project work
    b. Identifying risks
    c. Analyzing assumptions
    d. Creating a Responsibility Assignment Matrix

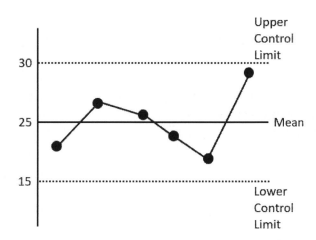

146. For the type of chart above, one standard deviation is ___ above and below the mean and two standard deviations is everything within ___ of the same mean.

    a. 34%, 47.2%
    b. 20%, 40%
    c. 28%, 42.5%
    d. 40%, 60%

147. Which of the following is not a project activity beginning during the project execution phase?

   a. Relying on checklists, the Delphi technique, and other strategies to handle risks
   b. Norming project work
   c. Under taking multi-criteria decision analysis
   d. The project manager establishing a consistent leadership style

148. Which of the following is not a project activity beginning during the project controlling and monitoring phase?

   a. Performance appraisals
   b. Rebaselining
   c. Variable sampling
   d. Relying on integrated project management software

149. Which of the following is not a project activity during the project closing phase?

   a. Product verification
   b. Producing project closure plans
   c. Procurement closure
   d. Writing a lessons learned report

150. Who are the two groups with whom a project manager should prioritize consistent communication?

   a. Project drivers and key stakeholders
   b. Project drivers and contractors
   c. Key stakeholders and functional managers
   d. Contractors and functional managers

151. If your organization already has a business in mind for procurement and skips or expedites the process of requesting and reviewing, this is called:

   a. Prerequisite
   b. Preassignment
   c. Prebidding
   d. Prerequesting

152. A process by which a project manager lets the project team know what metrics they will be checking to measure progress the next time the project is reviewed is called a:

a. Performance appraisal
b. Progress review
c. Forward view
d. Intermediate step

153. The project steering group is made up of:

a. Sponsor, project users, and project suppliers
b. Sponsor, project users, and project administration
c. Project users, project suppliers, and project managers
d. Project managers, team leaders, and project administration

154. What is the primary role of a project expeditor?

a. Support staff and coordinate communication
b. Determine and delegate ideal tasks to different team members
c. Authorize the project and provide guidance to the team
d. Facilitate interaction and productive communication

155. Which of these inputs is NOT a subsidiary plan?

a. Scope management
b. Risk management
c. Activity list
d. Procurement

156. Float is also known as:

a. Buffer
b. Leveling
c. Drift
d. None of the above

157. Analogous estimating is:

   a. A form of top-down estimating
   b. A form of bottom-up estimating
   c. A form of parametric estimating
   d. It depends on the project

158. The total amount needed to sustain all of a project team's resources, from electric bills to every salary is called:

   a. Sunk cost
   b. Budget baseline
   c. Full burdened rate
   d. Budgeted cost of work scheduled

159. Which of the following can be used to attain procurements?

   a. Invitation to bid
   b. Requests for information
   c. Requests for proposal
   d. All of the above

160. Your firm has been chosen by a new client because the client did not see the profit estimate they anticipated from a prior project conducted by a different firm. The client feels that the previous firm did not do enough to ensure that the product would be profitable. Which of the following should your firm conduct to ensure the client's needs are compatible with the firm's goals before taking on the client?

   a. Capital appropriations plan
   b. Legal contract
   c. Cost-benefit analysis
   d. Feasibility statement

161. A work flow diagram shows deliverables in the order:

   a. Of importance
   b. In which they must be completed
   c. In the order of which they must be started
   d. It depends on the type of work flow diagram

162. While configuration management coordinates the rollout of approved changes to the activities and goals of the project, more substantial changes to project plans, or whenever a new project plan is adopted, needs _____ to determine remaining project tasks and offer new quality checks.

   a. Critical path change
   b. Rebaselining
   c. Forward projections
   d. Control processes

163. In conflict management, what is the difference between smoothing and withdrawing?

   a. Smoothing involves a project manager stepping in to "smooth" out issues between team members by introducing a new set of guidelines to permanently address underlying issues, while withdrawing involves a project manager creating breathing space between the two parties in the short-term to see if the issue blows over.
   b. Smoothing involves a project manager creating breathing space between the two parties in the short-term to see if the issue blows over, while withdrawing implies that the project manager ignores rather than addresses the conflict.
   c. Smoothing involves a project manager creating breathing space between the two parties in the short-term to see if the issue blows over, while withdrawing involves a project manager taking a step back or "withdrawing" from the conflict to allow team members to develop a new set of guidelines to permanently address underlying issue.
   d. Smoothing involves a project manager ignoring rather than addressing the conflict, while withdrawing involves the project manager creating breathing space between the two parties in the short-term to see if the issue blows over.

164. Which of the following is an internal dependency?

   a. An event that occurs outside the project scope that triggers a schedule change
   b. A sequence of tasks not dictated by the nature of the work, but rather agreed to within the project team
   c. An event that occurs within the project team, but not tied to the execution of tasks, that can change the project schedule
   d. Tasks that have to be done in an exact order

165. Team involvement is most essential in:

    a. Work breakdown structure (WBS) creation
    b. Reserve analysis
    c. Making a viability decision
    d. Scope validation

166. An automotive industry project is under-budget and ahead of schedule. Team members' work efficiency is increasing. Project stakeholders are content and their feedback is positive. Project stakeholders are so satisfied that they are offering ideas for additional but modest features to add to the project. The project manager obliges and organizes work to realize stakeholders' suggestions.

The above scenario is an example of:

    a. Model project management
    b. Control scope
    c. Efficient communication
    d. Scope creep

167. If the schedule performance index (SPI) is 1.1 and the cost performance index (CPI) is 1, which of the following must be true?

    a. The project is behind schedule
    b. The project is right on schedule
    c. Planned value is greater than actual cost
    d. Planned value is less than actual cost

168. Network diagrams are meant to display:

    a. Critical paths
    b. Dependencies
    c. Relationships in-between tasks
    d. All of the above

169. The amount of time an activity can be delayed without extending project timelines is:

  a. An activity's float
  b. A critical path
  c. Maximax criterion
  d. Resource crashing

170. Towards the middle of project work, the client who commissioned the project doubles in size due to a previous investment paying off. The larger firm now has much more money to spend on the project. Instead of just continuing the project, however, the client offers to buy the performing organization and make the project work a permanent part of their business. The performing organization rejected the bid. In a second offer, the client asked the project to be merged with another two projects. The performing organization accepted and merged projects, thereby ending the original project. The larger project was successfully and fully completed.

Which of the following correctly describes the scenario above?

  a. After an integration proposal was rejected, an addition offer was accepted
  b. After an addition proposal was rejected, an integration offer was accepted
  c. After one addition proposal was rejected, another addition offer was accepted
  d. After one integration proposal was rejected, an integration offer was accepted

171. A project manager is facing a make-or-break point in project work. After a number of close calls, progress has slowed and the next set of tasks are crucial. If the next set of tasks cannot be completed on time, the project may be starved. All of these imminent tasks are dependent on one another such that one activity can only be completed when another activity is also completed. In a precedence diagram, these tasks have which type of relationship?

  a. Finish to start (FS)
  b. Finish to finish (FF)
  c. Start to start (SS)
  d. Start to finish (SF)

172. Which of the following is an example of a fixed cost?

   a. Project startup costs
   b. Reserve funds
   c. Administrative costs
   d. Bottom-up costs

173. What is a work breakdown structure?

   a. The work breakdown structure is a framework for understanding how part of a project affects the whole
   b. The work breakdown structure is the total amount needed to sustain all of a project team's resources, from electric bills to every salary
   c. The work breakdown structure is a chart that ties deliverables or activities to individuals on the project team to provide clear lines of responsibilities
   d. The work breakdown structure is a deliverable based breakdown of a project into lesser components

174. Which of the following phases does NOT apply to project life cycles?

   a. Initiation
   b. Execution
   c. Milestone
   d. Closure

175. Which of the following theories referring to cost of quality (COQ) focuses on output and believes an optimum number of allowable defects rounds out to 3.4 per 1,000,000?

   a. The plan-do-study-act (PDSA) cycle
   b. Total quality management (TQM)
   c. The Deming cycle
   d. The six sigma (6σ) measurements

176. You are a project manager for a product that was due in the third fiscal quarter. However, your client wants the project to be delivered in the second quarter. The client is willing to invest in more resources to ensure the project is completed sooner. Which of the following strategies should you employ first to get the entire project delivered at a quicker pace?

    a. Reverse resource allocation
    b. Resource crashing
    c. Resource leveling
    d. Fast tracking

177. What is the difference between total float and free float?

    a. Total float refers to the amounts of time an activity can be postponed without delaying the start of another separate activity, while free float refers to the amount of time an activity can be postponed and begun later without pushing back the overall project end date
    b. Total float refers to the amount of time an activity can be postponed and begun later without pushing back the overall project end date, while free float refers to the amounts of time an activity can be postponed without delaying the start of another separate activity
    c. Total float modifies a relationship between project activities and accelerates work on a dependent successor task when the preceding task is bogged down, while free float extends the time between two dependent tasks
    d. Total float extends the time between two dependent tasks, while free float modifies a relationship between project activities and accelerates work on a dependent successor task when the preceding task is bogged down

178. What is another term for soft logic?

    a. External dependency
    b. Discretionary dependency
    c. Mandatory dependency
    d. Internal dependency

179. A project manager's budget plan does not account for the varying exchange rate between the United States dollar and the United Kingdom's pound, where the product will be produced. The project manager did not adequately consider which of the following concepts when budget planning?

a. Opportunity cost
b. Economic profit
c. Level of accuracy
d. Cultural sensitivities

180. Which of the following is least important when developing a procurement plan?

a. Vendor processes
b. Timing
c. Roles and responsibilities
d. Lines of communication

181. The majority of a project manager's time is taken up with:

a. Planning
b. Communicating
c. Budgeting
d. Quality control

182. A _____ is the most detailed day-to-day description of project work.

a. Project activity
b. Project life cycle
c. Work breakdown structure
d. Work package

183. Which of the following is a characteristic of adequate project scoping?

a. A traceability matrix is developed to allow understanding of how deadlines were decided
b. Project assumptions are hypothesized before the project develops
c. The project is broken down to the work package level to anticipate workloads
d. The sunk cost should be determined before a legal contract is drawn up

184. A summary activity is also known as a:

a. Hammock
b. Portfolio
c. Milestone
d. None of the above

185. A project manager at a project-based firm is deciding whether to recommend a project to redesign a product for a prospective client. The net present value yields a negative value but the president of the firm and many stakeholders think the prospective client could introduce them to similar clientele if the product redesign is successful. What should the project manager do?

a. Recommend against the project for the time being
b. Recommend for the project
c. Present the NPV alongside a separate report on the value of the product's redesign, hear feedback, and then decide on a recommendation
d. Let the board or any other entity decide whether to proceed or not

186. Project managers use earned value measurement to:

a. Remain current with budgetary and spending plans
b. Get objective information on project work, free from functional managers' biases
c. Get up-to-date reports on project performance
d. All of the above

187. A large, multi-year marketing project is subject to quarterly audits. These audits are supposed to be all-encompassing but usually focus on just the financial aspects of the project. Only two weeks after the most previous audit, the project manager and team find out that an extensive quality-audit is to be undertaken. There are multiple complaints amongst the project team and opposition to this unannounced quality audit. Why is it important to cooperate fully with this new audit?

a. Achieving a high level of quality is the overriding purpose of a project
b. Having an outside source look for efficiency and effectiveness shortcomings will streamline work and benefit all parties
c. The project team is to execute not oppose project demands
d. A quality-audit can ensure the project is complying with industrial standards

188. Which of the following correctly identifies the key differences between a histograms and Ishikawa Diagrams?

   a. Histograms are used to display occurrences of an issue, while Ishikawa Diagrams try to find the origins of project problems
   b. Histograms are used to show the history of an issue, while Ishikawa Diagrams try to find the origins of project problems
   c. Histograms are used to display occurrences of an issue, while Ishikawa Diagrams try to find problems before they occur
   d. Histograms are used to display occurrences of an issue, while Ishikawa Diagrams try to find the cost of project problems

189. A critical part of a lessons learned report is the _____, which analyzes the strengths and weaknesses of how changes to the project plans were handled in the project.

   a. Final budget
   b. Schedule performance
   c. Project logs
   d. Controls review

190. When identifying stakeholder inputs, which of the following is an environmental factor you may want to pay notice to?

   a. Organizational structure
   b. Company culture
   c. Governmental/industry standards
   d. All of the above

191. Instead of traditional written reports, team leaders or project managers can use digital _____, such as work flow diagrams, to show delivery.

   a. Gauge reporting
   b. Presentations
   c. Dashboard reporting
   d. Demos

192. As a project manager, you need to help your company decide between taking two projects. Both projects will take place during a period of six months and both will generate a cash inflow of $75,000. However, Project A requires 50 hours of billable work per week, while Project B requires 40 hours of billable work per week. Which project would you advise stakeholders to take, ceteris paribus?

   a. Project A
   b. Project B
   c. They both have the same cash inflow per year so both are equally advisable
   d. Not enough information to make a decision

193. You are taking over from a project manager moved to another department early on in the initiation phase. The departing project manager tells you that she was working on the project's objectives, the desired deliverables, and anticipated timelines before the firm commits to the project. The project manager was working on which of the following?

   a. The subsidiary plan
   b. The project management plan
   c. The project charter
   d. The preliminary project scope statement

194. A linear responsibility chart:

   a. Is mostly drawn up for industrial or manufacturing projects in which minor physical defects could set the project back
   b. Displays the relationships between project activities
   c. Ties deliverables or activities to individuals on the project team to provide clear lines of responsibilities
   d. Has a vertical axis representing time and horizontal bars reflecting how long a project task is projected to take

195. Which of the following is NOT an input that would be relevant to estimating costs?

   a. Human resource management plan
   b. Scope baseline
   c. Core competencies
   d. Risk register

196. Rolling wave planning shapes the work breakdown structure (WBS) in what way?

    a.   The first WBS has subproject placeholders instead of specific timelines for project completion
    b.   Milestones and targets aren't detailed until the project evolves over time
    c.   Some deliverables cannot be mapped out at the start of a project, so the WBS is not completed at a single point
    d.   All of the above

197. One way Monte Carlo analysis can be used is to:

    a.   Indicate resource needs
    b.   Assess project risk probabilities
    c.   Address quality
    d.   Indicate which team-member is least efficient

198. A project is being considered that requires the involvement of a social media marketer who frequently makes spelling and grammar errors and is sensitive to criticism. During the initiation stage, how should a project manager expect to improve this social media marketer's performance?

    a.   Plan on either decreasing or increasing the social media marketer's future pay based on performance
    b.   Clearly state the social media marketer's duties ahead of time
    c.   Get the social media marketer's functional manager involved early on and ensure the manager understands the scope of the project
    d.   Get the social media marketer's functional manager involved early on and ensure the manager evaluates the social media marketer's performance based on project performance

199. During an information technology project, problems arise about quality and grade. During a meeting, a subject-matter expert asks: How are grade and quality different? How would you answer?

    a.   Grade measures condition, while quality measures features
    b.   Grade measures features, while quality measures condition
    c.   Grade and quality measure the same things but in different ways
    d.   Grade and quality are synonyms

200. A client that your firm has a positive history with praises your firm's handle on project processes and avoiding setbacks. They ask for you to eliminate project management costs since project controls inflate costs "unnecessarily" from their point of view. If the answers below are all possible or true, which of the following answers is the best way to respond to this client?

    a. Agree with your client and remove project management costs
    b. Remove some costs, such as those associated with communications and meetings
    c. Remove all costs but the project manager's salary
    d. Explain the costs sustained on previous projects that did not use project management

Practice exam answer key

1. C	2. A	3. D	4. D	5. B	6. B	7. C	8. C
9. A	10. C	11. D	12. D	13. D	14. A	15. C	16. A
17. B	18. C	19. A	20. A	21. A	22. D	23. B	24. A
25. A	26. D	27. C	28. B	29. C	30. A	31. D	32. C
33. A	34. D	35. D	36. A	37. A	38. B	39. C	40. C
41. C	42. D	43. C	44. A	45. C	46. A	47. A	48. C
49. C	50. D	51. A	52. A	53. C	54. D	55. B	56. D
57. D	58. B	59. B	60. D	61. A	62. B	63. D	64. A
65. D	66. C	67. C	68. B	69. D	70. D	71. D	72. D
73. A	74. A	75. B	76. A	77. D	78. B	79. C	80. C
81. C	82. C	83. B	84. C	85. B	86. A	87. B	88. C
89. A	90. C	91. A	92. A	93. B	94. D	95. B	96. D
97. C	98. A	99. B	100. D	101. B	102. D	103. B	104. B
105. C	106. A	107. D	108. B	109. A	110. B	111. A	112. D
113. B	114. D	115. D	116. B	117. A	118. D	119. B	120. B
121. D	122. C	123. D	124. C	125. B	126. A	127. B	128. B
129. B	130. D	131. B	132. C	133. A	134. D	135. C	136. C
137. B	138. C	139. D	140. D	141. C	142. A	143. A	144. C
145. A	146. A	147. D	148. D	149. B	150. A	151. B	152. C
153. A	154. A	155. C	156. D	157. A	158. C	159. D	160. D
161. B	162. B	163. B	164. C	165. A	166. D	167. D	168. D
169. A	170. B	171. B	172. A	173. D	174. C	175. D	176. B
177. B	178. B	179. C	180. D	181. B	182. D	183. C	184. A
185. A	186. D	187. B	188. A	189. D	190. D	191. C	192. B
193. D	194. C	195. C	196. D	197. B	198. D	199. B	200. D

1. Which of the following is an example of qualitative risk analysis?

   a. Calculating the damages various risks could inflict
   b. Gathering all of the possible risk events a project faces
   c. Prioritizing risks by potential damage
   d. Planning on ways to avoid risks

Answer: C. Qualitative risk analysis identifies risks and prioritizes risks by the effect they can have on the project. It is often a first step before a more detailed, quantitative analysis begins. Answer A is an example of quantitative risk analysis, B describes a risk register, and D is in a planning process instead of an analysis one.

2. The largest portion of a project's budget is typically spent in which phase?

   a. The execution phase
   b. The closing phase
   c. The monitoring and controlling phase
   d. The planning phase

Answer: A. The execution phase encompasses most of the work done for a project and therefore captures the greatest share of the budget. The monitoring and controlling phase overlaps with the execution phase. But the budget allocations for monitoring and controlling are distinct and typically much less than what is set aside for execution.

3. Within an organization that is managing a portfolio of projects that share resources, multiple projects are running behind. A project manager for one of these projects is growing frustrated with her project being ignored and resources taken away. Which of the following could help relieve this situation?

   a. Resource levelling
   b. Updating team communication protocols
   c. Appeals to higher management
   d. Contingency planning

Answer: D. Contingency plans are meant to help projects that are having issues with resources and deadlines and are appropriate here. Resource levelling is not the issue here as there is a competition for resources between projects and not within one project. The other two options should only be considered after all contingency plans have been tried.

4. During the monitoring and controlling phase, what actions should a project manager take to address the issue of a team member not responding to emails?

   a. Plan on either decreasing or increasing the team member's future pay based on performance
   b. Clearly state the team member's duties ahead of time
   c. Ensure the team member's functional manager understands the scope of the project
   d. None of the above

Answer: D. Answer A is not something a project manager can address. It is someone else's responsibility and not something that would fix the issue during the monitoring and controlling phase. Answers B and C are done during an earlier phase and not appropriate this late in the project. The project manager should schedule a meeting to discuss communication requirements. If that is resisted, one of the many conflict resolution strategies should be used.

5. Which of the following requires a project manager to seek a legal opinion?

   a. A stakeholder is discovered to have a conflict of interest with work done for one project deliverable
   b. An outside contractor requests access to proprietary information from the project manager's firm which had hitherto not been shared
   c. A project team member claims emotional trauma from overtime work
   d. All of the above

Answer: B. While all of these choices seem to need a legal input, only choice B requires it. A stakeholder's conflict of interest can be worked around and it is only a legal issue for projects that interface with government or organizations that require disclosure. It is not clear those conditions are met here. Human resource plans address members who are having difficulty with work schedules.

6. Which of the following statements is true?

    a. Programs are groups of related deliverables
    b. Portfolios group projects together for a common strategic, long-term reason
    c. Hammocks group together overlapping project phases
    d. Project life cycles consist of overlapping project milestones

Answer: B. Portfolios have multiple projects either being managed or considered. These projects are undertaken concurrently to achieve a goal. Programs are groups of related projects that are initiated simultaneously to achieve efficiencies. Hammocks look to find related activities that can be grouped together to streamline later project work. The project life cycle consists of overlapping project phases.

7. The risk register, human resource management plan, scope baseline, and cost management plan are all required in which project planning activity?

    a. Developing the schedule
    b. Defining activities
    c. Estimating costs
    d. Estimating activity duration

Answer: C. The listed inputs are directly relevant to estimating costs and none of the other planning activities. The risk register is a list of risks ordered by priority and including each risk's probability, likely cost, and timeline. Risk registers are diagramed, often as scatterplots, which try to ensure the budget accounts for project risks. The project schedule can also be considered for estimating costs.

8. A project manager has identified twelve risks for a deliverable and categorized them in a risk register, involved the relevant stakeholders, developed ways to track the risks, and rated them by priority. What would be the logical next step in the project manager's analysis?

    a. Negotiate higher risk thresholds
    b. Smooth out conflicts between parties that share risk
    c. Identify the triggers for each risk
    d. Determine if the risk is insurable or not

Answer: C. Triggers or risk symptoms are indications that a risk will materialize into a threat. Since the project manager in this scenario is clearly at the stage of evaluating risks that have yet to occur, answer C is the only appropriate answer. The other choices should be done only if necessary later during project work.

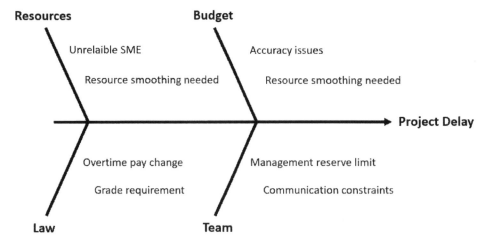

Resources

Budget

Unrelaible SME

Accuracy issues

Resource smoothing needed

Resource smoothing needed

Project Delay

Overtime pay change

Management reserve limit

Grade requirement

Communication constraints

Law

Team

9. This diagram's purpose is to:

a. Find the origins of project problems
b. Reflect how long a project task is projected to take
c. Display the relationships between project activities
d. Provide network analysis for complex projects

Answer: A. The diagram shown is a cause and effect diagram, also referred to as a fishbone or Ishikawa diagram. A cause and effect diagram's purpose is to find the origins of project problems. Answer B refers to a Gantt chart, answer C refers to a PERT chart, and answer D refers to a GERT chart.

10. Your team member is two days late with a report. Just before a meeting in which the report's topic will be the main subject discussed, he emails it to you. You immediately notice that the report has serious errors. What should you do?

a. Hold the meeting and allow the team member to present the report, errors and all
b. Hold the meeting and inform your team members there are errors in the report
c. Delay the meeting until the report has been corrected by the team member
d. Delay the meeting and correct the report yourself

Answer: C. The meeting should be rescheduled. Proceeding with the meeting and pointing out the errors only serves to embarrass the report's writer. Presenting inaccurate information wastes team members' time. Let the team member know the issues with the report, but do not fix the report yourself – this decreases their ownership and wastes your time. Throughout the project, the project manager should enforce team responsibilities. Escalate the issue if it persists.

11. A project statement of work (SOW) includes which of the following?

    a.   The summary activity, a list of all deliverables, and strategic vision behind the project
    b.   A list of all deliverables and the project scope
    c.   A list of all deliverables, list of all team members, and the project scope
    d.   The project scope and the strategic vision behind the project

Answer: D. A SOW summarizes at a high level. It should include an analysis of the project scope and how the project fits into a firm's strategic goals. The other answer choices include options that are too specific. The summary activity choice in answer A is a non sequitur meant to throw you off.

12. Who has access to the stakeholder register?

    a.   Only the project manager
    b.   The listed stakeholders
    c.   All team members and stakeholders
    d.   Anyone with whom the project manager shares the register

Answer: D. The stakeholder register should be available to individuals only at the project manager's discretion. Since the stakeholder register includes sensitive information such as the project manager's impressions of stakeholders (including their skills, attitude, or potential challenges to working with them), the stakeholder register shouldn't be available for all to see. However, there may be times when a project manager may feel it is appropriate or necessary to share the register with others.

13. What are the benefits of a RACI chart?

    a.   It lowers risk by making responsibilities clear
    b.   It shows which tasks team members spend time on, even if it is a small amount
    c.   It shows the roles of the project team in various project activities
    d.   All of the above

Answer: D. RACI is an acronym that represents the process of determining who is responsible, accountable, consulted, and informed per each project activity. By ensuring responsibilities are clear for all team members, a RACI chart lowers the risks that can occur from lack of understanding or communication and can serve as a way to track which tasks have been completed by which team members, as well as their performance in completing them.

14. A code of accounts assists which aspect of project managing?

    a. Time management because a code of accounts simplifies jargon in communication
    b. Hammocking because a code of accounts is ordered in a way to logically sequence tasks
    c. Time management because a code of accounts simplifies cost estimates
    d. Decomposition because a code of accounts allows an easy way to breakdown deliverables

Answer: A. A code of accounts is a customized numbering or lettering system that allows for a shortening of complex jargon or product identifiers. Instead of memorizing many different codes, a code of accounts provides a simple reference point.

15. How do building information modeling (BIM) and project information systems (PMIS) relate?

    a. Both refer to online collaborative environments meant to increase team cohesiveness and are launched during the execution phase of a project
    b. Configuration management is a part of BIM and PMIS is focused on dynamic scheduling
    c. Both refer to online collaborative environments meant to increase team cohesiveness and are launched during the initiation phase of a project
    d. BIM is focused on dynamic scheduling and PMIS is focused on stakeholder engagement

Answer: C. PMIS is an online database created for the shared use of project team. The PMIS enables project managers to organize and disseminate progress reports on each facet of a project. There are specialty PMIS for specific types of projects. Building information modeling (BIM) is a common PMIS used in the construction industry. Since PMIS are used in the planning phase, they are launched at the end of the initiation phase.

16. You are 40% of the way done with a project to upgrade a bank's internal network. There are 100 locations in Canada and another 50 in the United States. An accounting software vendor has just released a significant software upgrade for most of the equipment you are installing. This software upgrade would offer the customer increased functionality that they initially said they needed that did not exist when the project began. What should you tell the bank?

    a. Let the bank know about the upgrade and its potential impact on the project's timeline and functionality if implemented
    b. Install the upgrade and change the timeline as necessary, since the bank said they needed this type of functionality to begin with
    c. Install the upgrade on the remaining sites and keep your timeline on schedule
    d. Tell them nothing and continue along the same trajectory. The bank did not anticipate the update

Answer: A. A project manager should always look after their client's best interests. Not informing the bank about the newly available upgrade would be against their best interests and contrary to project management ethics. At the same time, simply upgrading the software without informing them of the probable schedule changes or only partially upgrading their software would be equally irresponsible as the scope has already been approved and changes should go through all necessary change control processes.

17. You have recently been promoted to project manager because the last project with a specific client resulted in dissatisfaction and higher costs than expected. This new project requires you to create a system that ensures a manufacturer's products satisfy consumer safety laws. The manufacturer expects this system to quickly reduce his cost. Which of the following is not true in this situation?

    a. Assumptions about consumer safety laws should be evaluated to ensure the system effectively meets standards
    b. The objective is to do better than the last project manager
    c. The schedule is a constraint partly because the owner expects a quick result
    d. The budget is a constraint partly because the last project came in over budget

Answer: B. The objective is to meet the project's goals. Rivalries or past failures should not be centered in project work. Lessons learned from past failures are important, but objectives always revolve around the clients' needs and project work. When working with standards, assumptions should always be evaluated.

18. A project manager leads a team of twenty people. Some individuals communicate daily, and others are silent for days at a time. More often than not, the project manager is carbon-copied on emails. How many lines of communication exist in this project?

 a. 10
 b. 20
 c. 105
 d. 210

Answer: C. Lines of communication are determined by the number of people communicating, not the frequency of communication. The equation to use is:

$$lines\ of\ communication = \frac{n(n-1)}{2} = \frac{15(14)}{2} = \frac{210}{2} = 105$$

19. Which of the following statements about change control is true?

 a. The change control team should first record and categorize client requests for a change
 b. Fixed-price contracts make change controls less necessary
 c. Regression plans should be created during the assessment phase of change control
 d. Change control is the same thing as configuration management

Answer: A. Recording or classifying a client's change request is the first step of most change control procedures. All of the steps are: Record, Assess, Plan, Build, Implement, and Gain Acceptance. While a fixed-price contract might reduce the need for changes, it would not eliminate the causes of changes. Regression plans should be constructed during the "plan" phase of the change control. Configuration management, while similar to change control, concentrates on the specifications of deliverables and processes, while change control is more specific in that it focuses more on controlling the changes to the project and its project baselines.

20. What is the difference between regression analysis cost estimating and parametric cost estimating?

    a. Regression analysis cost estimating looks for statistical trends between the costs of similar deliverables to estimate costs at varying scales. Parametric cost estimating uses mathematical formulas to estimate costs of a deliverable by using the relationships between variables like per unit cost.

    b. Parametric cost estimating looks for statistical trends between the costs of similar deliverables to estimate costs at varying scales. Regression analysis cost estimating uses mathematical formulas to estimate costs of a deliverable by using the relationships between variables like per unit cost.

    c. Regression analysis cost estimating is the average of the likely, worst-case, and best-case cost estimates. Parametric cost estimating examines the budgets of similar completed projects within the firms or accessible budgets (for example, government contractors) from outside the firm to estimate costs.

    d. Parametric cost estimating is the average of the likely, worst-case, and best-case cost estimates. Regression analysis cost estimating examines the budgets of similar completed projects within the firms or accessible budgets (for example, government contractors) from outside the firm to estimate costs.

Answer: A. Regression analysis cost estimating looks for statistical trends between the costs of similar deliverables to estimate costs at varying scales. Parametric cost estimating uses mathematical formulas to estimate costs of a deliverable by using the relationships between variables like per unit cost. Answers C and D refer to three point estimating, which is the average of the likely, worst-case, and best-case cost estimates, and analogous cost estimating, which examines the budgets of similar completed projects within the firms or accessible budgets (for example, government contractors) from outside the firm to estimate costs.

21. Which of the following is not a characteristic of adequate project scoping?

   a. A requirements traceability matrix is developed to allow understanding of how deadlines were decided
   b. Project assumptions are listed and tested as the project develops
   c. The project is broken down into work package level to anticipate workloads
   d. Measurement criteria for many project requirements are determined

Answer: A. A traceability matrix allows for many things to be traced, such as measurement criteria and stakeholder input. Deadlines should be spelled out in the contract or decided by the project manager and do not need to be traced to a nonobvious source. Traceability is concerned with work validation, not project scoping.

22. During the controlling of project work, a project manager submits a report on project progress to other contractors to review. As the report circulates for further review, the project manager notices that data he entered earlier had been altered at multiple places. The data is altered to overstate project progress and outside contractors were not involved in the altered areas. What action should the project manager take?

   a. Reenter the data to be accurate and only take further action if it is altered again
   b. Email the contractor most likely to have altered the data to see if they did it
   c. Email every contractor to discover who altered the data
   d. Notify higher management to examine the issue

Answer: D. The ethics of project management require that a project manager notify others of a breach of project trust and fidelity. Especially with outside partners, a project manager should involve those with more authority to deal directly with contractors' higher management.

23. Which of the following statements is false?

   a. Programs are groups of related projects managed together
   b. Portfolios are groups related deliverables
   c. Hammocks group together overlapping project tasks
   d. Project life cycles consist of overlapping project phases

Answer: B. Portfolio management is management of all an organization's projects from a high-level perspective.

24. Both simple average and PERT calculations can help determine:

   a. Three-point estimation
   b. Project scheduling
   c. Sequencing activities
   d. Estimating activity durations

Answer: A. While three-point estimation is developed by considering the most pessimistic, optimistic and most-likely outcomes, similar outcomes can be developed by using the simple average or PERT average formulas. Simple average and PERT are more common ways of arriving at three-point estimations.

25. During the planning phase, a functional manager cleared ten team-members for project work three months down the line. When the time arrives, the functional manager only provides five workers and says her team will need double the scheduled time. Which of the following outlines the appropriate response?

   a. If the relevant work is on a critical path, evoke contingency plans and resource reserves to find ways to prevent work delays
   b. If the relevant work is on a critical path, appeal to higher management to get more resources
   c. If the relevant work is not on a critical path, evoke contingency plans and resource reserves to find ways to prevent work delays
   d. If the relevant work is not on a critical path, appeal to higher management to get more resources

Answer: A. Critical path work delays will have ripple effects and throw many other project activities off. If the relevant work is not on a critical path, float time should be examined. If the delay exceeds the float, then contingency plans should be evoked. Answer C did not examine float time and is incorrect.

26. In which of the following organizational forms does project management hold less sway than expertise?

   a. Matrix
   b. Laissez faire
   c. Type X
   d. Functional

Answer: D. A functional form during a project empowers functional managers, those with specialized skillsets, over project managers. Project managers in such a setup have limited authority to set timelines and large project priorities.

27. What is the chief difference between the program evaluation review technique (PERT) and the critical path method (CPM) charts?

    a.   CPM prioritizes dependencies between tasks while PERT prioritizes duration between tasks
    b.   PERT prioritizes dependencies between tasks while CPM prioritizes duration between tasks
    c.   CPM uses most likely duration while PERT uses mean of estimates
    d.   PERT uses expected value while CPM uses most likely duration

Answer: C. The two are similar in many ways. Both focus on dependencies and duration. The chief difference has to do with how they calculate duration, which is what answers C and D focus on. Expected value is a probabilistic calculation that uses three-point estimates for each activity and then applies the average to the entire project equally. CPM does not use expected value and merely uses the mean, avoiding worst- and best-case scenarios.

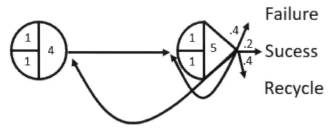

28. The numbers in this chart come from:
    a.   Probability multiplied by impact
    b.   Running mathematical Monte Carlo simulations
    c.   Determining the number of nodes of communication
    d.   Subtracting the probable expected loss from the probable expected profit

Answer: B. The chart in this question is graphic evaluation and review technique (GERT) chart, which is typically used for analysis for complex projects, often in engineering. Running mathematical Monte Carlo simulations provides the numbers in a GERT chart. These simulations randomly sample outcomes using random methods to try to quantify the totality of possible outcomes. Answer A is the equation for determining expected value. For answer B, while nodes are a part of GERT charts, determining the number of nodes of communication refers to the lines of communication a project manager has. Answer D refers to expected monetary value (EMV) of an opportunity or risk.

29. In which of the following contracts is the profit unknown?
   a. Cost plus contract
   b. Cost reimbursable contract
   c. Fixed price contract
   d. Time and materials contract

Answer: C. From the perspective of the buyer, the profit is unknown in a fixed price contract.

30. A nanotechnology firm needs to use one of the world's fastest supercomputers during one phase of an upcoming project. The supercomputer charges by the hour. The project manager works backwards from the project's project completion date to best schedule the time when the supercomputer will be used. The project manager is employing which of the following resource-scheduling techniques?

   a. Reverse resource allocation
   b. Resource crashing
   c. Resource leveling
   d. Resource smoothing

Answer: A. Anytime a resource is only used once and is dependent on other activities, reverse resource allocation should be used to schedule the resource's use. A project manager works backwards because task dependencies can be more easily checked from the end date.

31. What is the difference between the percent-complete technique and the milestone technique?

a. The percent-complete technique calculates that project activity is zero percent before relevant work begins, fifty percent while work is ongoing, and one hundred percent complete once work is finished, while the milestone technique measures project progress by the total amount of projective activities that are complete, with partially complete activities not counting towards the percent of project actually completed calculation.

b. The percent-complete technique measures project progress by the total amount of projective activities that are complete, with partially complete activities not counting towards the percent of project actually completed calculation, while the milestone technique estimates the exact amount of the project completed. This method requires the most meticulous project monitoring to not lead to an erroneous EV estimate.

c. The percent-complete technique measures project progress by the total amount of projective activities that are complete, with partially complete activities not counting towards the percent of project actually completed calculation, while the milestone technique calculates that project activity is zero percent before relevant work begins, fifty percent while work is ongoing, and one hundred percent complete once work is finished.

d. The percent-complete technique estimates the exact amount of the project completed. This method requires the most meticulous project monitoring to not lead to an erroneous EV estimate, while the milestone technique measures project progress by the total amount of projective activities that are complete, with partially complete activities not counting towards the percent of project actually completed calculation.

Answer: D. The percent-complete technique estimates the exact amount of the project completed. This method requires the most meticulous project monitoring to not lead to an erroneous EV estimate, while the milestone technique measures project progress by the total amount of projective activities that are complete, with partially complete activities not counting towards the percent of project actually completed calculation. Answers A and C refer to the fifty-fifty technique, which is a compromise between these two methods.

32. A project manager has just finished reviewing the mandatory relationships between project activities. The project manager was engaged in which of the following processes?

  a. Developing the schedule
  b. Defining activities
  c. Sequencing activities
  d. Estimating activity durations

Answer: C. Mandatory relationships express sequencing constraints. Certain activities can only be done in a certain order. Mandatory relationships are also termed hard logic.

33. Which of the following qualities is the least important quality of an effective project manager?

  a. Unfaltering self-reliance
  b. Changing plans after failures occur
  c. Productive and efficient communication
  d. Facilitating interaction

Answer: A. A project manager is at the center of a team-effort. The team is being managed in order for the project to be completed. All of the answers except A understand that reality. Self-reliance is great, but does not address the chief purpose of a project manager.

34. How does the procurement plan relate to other project plans?

  a. It makes up a large part of the budget and cost management plan
  b. It shapes the human resource plan by defining the level of expertise needed to work with various resources
  c. It can determine the schedule's sequencing of tasks by revealing where there are resource constraints
  d. All of the above

Answer: D. The procurement plan is a subsidiary plan that decides what resources will be bought, when they will be bought, and the vendors used. It affects teamwork, budget, and schedule planning in the ways outlined in the answer choices.

35. Clark is a project manager that oversees a team of eight that has worked together on more than a dozen projects. For their latest project, his team has determined that there are 235 risks and 12 major causes of those risks. The project work is nearly complete and has been approved by all key stakeholders, and the sponsor has been very supportive and involved. Despite all this, there is one risk the team cannot determine an effective way to prevent or insure against, and the work cannot be outsourced or removed. What should Clark do?

    a.   Determine a way to transfer the risk
    b.   Determine a way to avoid the risk
    c.   Determine a way to mitigate the risk
    d.   Accept the risk

Answer: D. Most of the initial information provided in this question is extraneous. The key is this sentence: "Despite all this, there is one risk the team cannot determine an effective way to prevent or insure against, and the work cannot be outsourced or removed." While it would be ideal if Clark and his team could determine a way to transfer, avoid, or mitigate the risk, that simply isn't possible. Therefore, Clark's only option is to accept the risk as one that the project team has to take.

36. All projects share which of the following characteristic(s)?

    a.   A start point and end point
    b.   Cyclically repeating activities
    c.   A project coordinator on staff
    d.   All of the above

Answer: A. All projects have clear beginning and end. Project activities are hard to generalize. Smaller projects may have few activities. Project coordinators are not necessary for a project.

37. Within a profit-maximizing firm, which of the following situations uses the opportunity cost concept best?

    a. During the budget-planning process, a project is estimated to total out at $300,000 and net a profit of $50,000. The project is taken because other possible projects yield a net less profit
    b. During the budget-planning process, a project is rejected because the project will yield a negative profit
    c. During the budget-planning process, the project manager compares costs of analogous projects to see which is the best opportunity
    d. During the budget-planning process, a firm considers all the ways to cut costs while maintaining quality

Answer: A. Figuring out opportunity cost requires looking at all possible opportunities, forgoing alternatives and choosing one. A firm must ask itself what else can we do without time instead and which option will lead us to our goal. Only answer A demonstrates that thought-process.

38. In engineering or IT projects, "configuration management" is a synonym for:

    a. Quality control
    b. Version control
    c. Quality assurance
    d. Project assurance

Answer: B. Configuration management is essentially version control or versioning. However, the term is typically only used in engineering or IT circles.

39. A "post-mortem" is used for a project's:

    a. Closure
    b. End
    c. Evaluation
    d. Completion

Answer: C. A post-mortem is often slang for a project's evaluation. A project's evaluation can also be known as a post project review (PPR). However, these two terms often carry negative connotations, so it is more appropriate to refer to this period in the project as its evaluation or lessons learned.

40. You are taking over from a project manager that moved to another department early on in the initiation phase. The departing project manager tells you that she was working on share strategies and triggers. The project manager was working on which of the following?

a. The cost management plan
b. The quality management plan
c. The risk management plan
d. The procurement management plan

Answer: C. Triggers or risk symptoms are indications that a risk will materialize into a threat. A share strategy is a form of risk transfer that deals with comparative advantage and turning risks into opportunities. It involves outsourcing risks to a third-party.

41. Jorge is a project manager who also happens to be a technical expert in his project's field, information technology. His project revolves around installing new software onto the computers of five separate departments of a Fortune 500 company. He also has a background in communications and management. The project has been going relatively well, and the software is resulting in many positive and necessary benefits for the company. However, a significant and unexpectedly high number of changes to project plans have been made as client feedback is analyzed. Which of the following reasons is the most likely cause of this issue?

a. The project needs additional management oversight since has so many benefits to the company
b. The project management team needs to use more project management communication processes
c. Some stakeholders were not identified by Jorge and his team
d. Jorge and his team do not fully understand the company's environment

Answer: C. The most likely cause of Jorge's issues is that some stakeholders were missed. Therefore, their requirements were not determined, and now they are requesting changes to accommodate these requirements. There is not enough evidence to think that additional management oversight, processes, or training about the company itself are the main contributing factors to these changes.

42. The chairperson of the project steering group is the:

a. Team leader
b. Project manager
c. Project user
d. Project sponsor

Answer: D. The chairperson should be the project sponsor. Project sponsors can also be known as the project director, project executive or simply executive, or a senior responsible owner (SRO).

43. The four types of cost-reimbursable contracts are:

a. CPFF, CPIF, CPAF, CPEC
b. CPFF, CPIF, CPOF, CPPC
c. CPFF, CPIF, CPAF, CPPC
d. CPEF, CPIF, CPAF, CPEC

Answer: C. The four types of cost-reimbursable contracts are: Cost Plus Fixed Fee (CPFF), Cost Plus Incentive Fee (CPIF), Cost Plus Award Fee Contracts (CPAF), and Cost Plus Percentage Cost (CPPC) Time and Materials (also known as Cost Plus Fee (CPF).

44. A project manager is overseeing project activities done consecutively and is nervous about the risks of planning later project activities before knowing more details. It is decided that the project plans will be finalized piece by piece, as project work sheds light on future requirements. Planning out project work in this manner is known as which of the following?

a. Rolling wave planning
b. Stage-gate process
c. Decomposition
d. Expert judgment

Answer: A. Rolling wave planning details project plans more over time—the project is done in waves. With a rolling wave approach, the project begins purposely with many unknowns that are only defined by subsequent project work. Answer B is the only other reasonable option. But the stage-gate process is used when there are risks about doing many activities at the same time.

45. Which of the following is NOT a part of the scope baseline?

    a. Project scope statement
    b. WBS
    c. Project charter
    d. Scope statement

Answer: C. The scope baseline, which is a part of the project management plan, consists of the project scope statement, WBS (work breakdown structure), and WBS dictionary (a guide to the WBS), the scope management plan, and the scope statement. The project charter may talk about scope but it is an artifact of the initiation phase and superseded by more precise information for the baseline.

46. Which of the following is not a grid that can be used to classify the power and influence of each stakeholder?

    a. Power/impact grid
    b. Influence/impact grid
    c. Power/influence grid
    d. Salience model

Answer: A. There are four classification models: the influence/impact grid, the power/influence grid, the power/interest grid, and the salience model.

47. What are the three main types of project budgets?

    a. Stage budget, risk budget, change budget
    b. Stage budget, change budget, project budget
    c. Maintenance budget, risk budget, change budget
    d. Stage budget, change budget, maintenance budget

Answer: A. The three project budgets are: stage budget (or the money for the planned work of the stage), change budget (or the money put aside to cover changes), and risk budget (or the money put aside for potential financial impacts from risks).

48. Which of the following is NOT one of the three costs associated with cost of quality (COQ)?

a. Prevention
b. Appraisal
c. Completion
d. Failure

Answer: C. The three costs associated with cost of quality (COQ) are prevention, appraisal, and failure (also called the cost of poor quality). These costs cover all the work needed to meet the product's quality requirements.

49. In the wake of an economic recession, the US government created the program Car Allowance Rebate System (CARS), more affectionately known as "Cash for Clunkers," in 2009. This program provided Americans with monetary incentives to trade in their less fuel-efficient vehicles for new, more fuel-efficient vehicles. If you were a project manager working for an automotive company back in 2009, which of the following methods would analyze the way CARS would change a project's forecast?

a. Time Series Methods
b. Judgmental Methods
c. Econometric Methods
d. None of the above

Answer: C. Econometric Methods, also known as Causal Methods, are used to identify variables that might change your project's forecast. Methods that are included in this group include econometrics, regression analysis, and autoregressive moving average (ARMA).

50. Which of the following could be included in a work performance report?

a. Time completion forecast
b. Risk and issue status
c. Results of variance analysis
d. All of the above

Answer: D. All of these factors could potentially be included in a work performance report. In reality, any information that stakeholders either want or need to know should be included in a work performance report.

51. Penelope has just started as a project manager for a large project that is half-way into its one year run. The project has 22 members on her project team, and also involves 10 sellers. In order to quickly appraise how the project is currently doing, which type of report should Penelope review?

a. Progress
b. Work status
c. Forecast
d. Executive

Answer: A. Since Penelope needs to "quickly" appraise the project's standing, a progress report would be her best bet since it summarizes a project's status. A work status report is typically more detailed, and a forecast report is for looking into the future, whereas an executive report is typically higher level and may not have the type of information she would need to know as a new project manager.

52. What is the difference between calculating cost variance and cost performance index?

a. Cost variance calculates the difference between the earned value and actual cost while cost performance index calculates the quotient of earned value to actual cost
b. Cost variance calculates the difference between the earned value and actual cost while cost performance index calculates the quotient of earned value to planned value
c. Cost variance calculates the quotient of earned value to planned value while cost performance index calculates the difference between the earned value and actual cost
d. Cost variance calculates the difference between the earned value and actual cost while cost performance index calculates the quotient of earned value to planned value

Answer: A. Cost variance (CV) calculates the difference between the earned value (EV) and actual cost (AC). The formula for CV is:
$$CV = EV - AC$$
Cost performance index (CPI) calculates the quotient of earned value (EV) to actual cost (AC). The equation for CPI is:
$$CPI = EV/AC$$
Answers C and D both have the schedule performance index (SPI) equation as an option. SPI calculates the quotient of earned value (EV) to planned value (PV). The equation for SPI is:
$$SPI = EV/PV$$

53. Anna is a first time project manager creating a procurement statement of work. A seller will be doing most of the actual work. Some stakeholders think that the procurement statement of work should only include the functional requirements while others want as many items as possible added. What would you advise Anna do in this situation?

    a. Determine if the seller has greater expertise than the buyer, and if so, only include functional requirements
    b. Determine if the seller has greater expertise than the buyer, and if so, be as detailed as possible
    c. Ensure the procurement statement of work is as detailed as needed for the project itself
    d. Keep the procurement statement of work as general as possible to allow for later clarification

Answer: C. Anna should ensure the procurement statement of work is as detailed as needed for the project itself. While the procurement statement of work can generally stick to function when the seller has more expertise than the buyer, it is still a best practice that a procurement statement of work should be as detailed as possible.

54. Which of the following is not a benefit of a RACI chart?

    a. It sets accountability for project activities
    b. It clearly demonstrates decision-making authorities
    c. It projects the roles of the project team in various project activities
    d. It illustrates the sequence and dependencies between various project activities

Answer: D. RACI is an acronym that represents the process of determining who is responsible, accountable, consulted, and informed per each project activity. Answer D describes the precedence diagram method.

55. Configuration Status Accounting refers to:

    a. Accounting for the status of funds for deliverables
    b. Accounting for the status of changes to the specifications of deliverables
    c. Accounting for the status of redirected funds for deliverables
    d. None of the above

Answer: B. Despite the word "accounting," configuration status accounting has to do not with accounting in the financial sense but in accounting for the status of changes to the specifications of deliverables. Anytime you see "configuration" in a question, you should look for change.

56. Which of the following is NOT one of the three steps of stakeholder analysis?

    a. Identify stakeholders
    b. Analyze potential impact
    c. Assess stakeholders' likely reactions
    d. Consider stakeholders' biases

Answer: D. The three steps of stakeholder analysis are: 1) identify stakeholders; 2) analyze potential impact; 3) assess stakeholders' likely reactions.

57. When should the project evaluation report be completed?

    a. Final delivery stage
    b. Every closure stage
    c. Project completion
    d. Project closure

Answer: D. The project evaluation report should be completed at the very end of the project, known as project closure when all of the project is evaluated. A project completion report is submitted at project completion.

58. The _____ is the collection of all records dealing with contracts attached to the project.

   a. Final review
   b. Procurement file
   c. Project postmortem
   d. Controls review

Answer: B. The procurement file is the collection of all records dealing with contracts attached to the project. The file should be referenced when deciding with whom to share the lessons learned report and to ensure that each of these relationships is reviewed in the report. Answer A refers to the last document disseminated to all parties associated with a project. Answer C, project postmortem, is another term for a lessons learned report. Answer D, the controls review analyzes the strengths and weaknesses of how changes to the project plans were handled in the project.

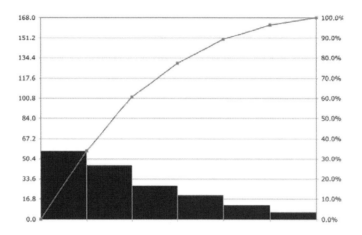

59. For a Pareto diagram like the one pictured above, the numbers on the left vertical axis represent _____, while on the right vertical axis the numbers represent _____.

    a. On the left vertical axis are cumulative measurements of each occurrence from left to right. On the right vertical axis are frequency percent measurements representing how often the phenomena on the horizontal axis appeared.

    b. On the left vertical axis are frequency measurements representing how often the phenomena on the horizontal axis appeared. On the right vertical axis are cumulative percent measurements of each occurrence from left to right.

    c. On the left vertical axis is the total amount needed to sustain all of a project team's resources. On the right vertical axis are cumulative percent measurements of how many funds have been spent from left to right.

    d. On the left vertical axis are frequency measurements representing the allotted amount of time in which a project can be completed in hours. On the right vertical axis are cumulative percent measurements of a project's completion.

Answer: B. Pareto diagrams have two distinct vertical axes. On the left vertical axis are frequency measurements representing how often the phenomena on the horizontal axis appeared. On the right vertical axis are cumulative percent measurements of each occurrence from left to right. The size of each bar and slope of the cumulative percent line help a team see which trends matter the most.

60. Many organizations create this type of board to authorize change requests:

a. Technical assessment board
b. Change control board
c. Configuration control board
d. All of the above

Answer: D. Depending on your organization's line of work, boards that authorize change requests can be called technical assessment boards, configuration control boards, or engineering review boards. However, the most common moniker for this type of board is change control board.

61. How is a configuration management plan different from a change management plan?

a. Configuration management plans deal with changes to parts of a product of the project, while change management plans deal with the project process
b. Configuration management plans deal with the project process, while change management plans deal with changes to parts of a product of the project
c. Configuration management plans deal with how you define, monitor, control and change the project schedule while change management plans deal with how you define, monitor, control and change human resources
d. Configuration management plans deal with how you define, monitor, control and change human resources while change management plans deal with how you define, monitor, control and change the project schedule

Answer: A. Configuration management plans deal with changes to parts of a product of the project. A change could be refining a product's physical characteristics or some of its features, for example. Change management plans, on the other hand, deal with the project process itself.

62. Stakeholders are most critical in which of the following project activities?

    a. Budget plans
    b. Writing a lessons learned report
    c. Procurement negotiations
    d. Conducting a feasibility study

Answer: B. Stakeholders are involved in every project activity to a degree. Any evaluation report written at the end of a project requires stakeholder input and surveying. Lessons learned revolve around stakeholder's experience and satisfaction. All the other answer choices have more limited stakeholder involvement and may not require stakeholders at all. For example, budget plans are internally focused and stakeholders may not offer any insight to better plan project spending.

63. Which of the following is a correct sequence of project areas to focus on, from earliest to latest?

    a. Project schedule, communication requirements, risk identification, and project budget
    b. Communication requirements, project budget, project schedule, and risk identification
    c. Project budget, project schedule, risk identification, and communication requirements
    d. Project schedule, project budget, communication requirements, and risk identification

Answer: D. The project schedule is the first plan that is attempted once a project is being seriously considered or initiated. The budget requires the schedule to make basic estimations about days of work that are needed. Once those two are completed, communication requirements are specified so work can commence. Risk identification cannot occur until project work is about to begin and the scope is fleshed out.

64. When does the initiation phase of a project end?

    a. When the project charter is signed by all parties
    b. When a project is being considered
    c. When project feasibility is validated
    d. As project monitoring begins

Answer: A. This question can be read as when does project planning begin, since after initiation comes the planning phase. The signing of the project party by all parties signifies the end of project initiation.

65. A formerly productive team-member has fallen behind. The quality of her work has also declined. Communication does not seem to be getting through to her. What would a theory z project manager prioritize to improve this situation?

    a. The consequences of not meeting project standards, such as getting fired
    b. The quality of team-member's past work and how valuable she is to the team
    c. Get higher management involved to prioritize the situation
    d. Improve the team-member's work environment

Answer: D. Answer A is a theory x response. Theory x managers are pessimistic about worker motivations and stern. Theory y managers are more optimistic and believe workers can be self-starters, as seen in B. Both focus on workers, while theory z focuses on the larger work setting. Answer C is an escalation, a conflict resolution technique and not a theory of management.

66. Project work that clarifies future work and enables project management to be more exacting is:

    a. Controlled
    b. Monitored
    c. Iterative
    d. Management by exception

Answer: C. Iterative project work is a combination of waterfall and agile approaches. All three of these models use lessons from prior work to improve future work and proceed in cycles of greater detail.

67. Project work is about half way finished and the project charter has been altered multiple times. It looks like the work to meet the next set of deliverables will be beyond the agreed-upon scope and require more charter alternations. Who has the ultimate responsibility to sign off on changes to the charter?

    a. The steering committees
    b. The project manager
    c. The project client
    d. The project manager's supervisors

Answer: C. The client pays for increased scope and so must agree to change the parameters of the project. The project manager, the steering committee, and the project manager's supervisors put together the change proposal and the decision to accept or not is the sponsor's.

68. Which of the following illustrates what W. Edwards Deming contributed to cost of quality (COQ) understanding?

    a. Total quality management
    b. Plan-do-check-act cycle
    c. Six sigma
    d. Continuous improvement

Answer: B. W. Edwards Deming is known for the plan-do-study-act (PDSA) cycle. Deming thought of quality as chiefly the responsibility of project managers. The PDSA tests a change and widely shares the results so a method can be improved. The PDSA is action-oriented and involves experimentation.

69. Which of the following might be a way of conducting a procurement performance review?

    a. Examining the quality of contract work
    b. Quality audit
    c. Inspection of documents
    d. All of the above

Answer: D. Procurement performance reviews can take many forms, such as examining the work of the product itself, performing a quality audit or the inspection of documents. They can also be formalized and conducted in intervals or at the end of the contract.

70. Any authorized change to a project plan is a scope change. On the other hand, _____ comprises all unauthorized changes to the project plan that are not authorized as they happen.

    a. Scope reworks
    b. Scope issues
    c. Scope intervention
    d. Scope creep

Answer: D. Scope creep comprises all unauthorized changes to the project plan that are not authorized as they happen. All extra work, no matter how small an addition to the work breakdown structure, constitutes scope creep. Options A, B and C are not terms used in project management.

71. When controlling for quality, taking a fraction of the total to draw conclusions about the total is called:

a. Forward projections
b. First unit
c. Rebaselining
d. Sampling

Answer: D. Sampling can be done for all projects and is another method to control for quality. Sampling means taking a fraction of the total to draw conclusions about the total. Answer A, forward projections, uses past project data, such as productivity and how far from reality project plans proved to be, to predict the future course of the project and the effects of all the changes that go into a new baseline. Answer B, first unit, is the first complete product the project produces before creating many more of them. Answer C, rebaselining, is less related to the latter stages of the project and is instead for more substantial changes to project plans when a new baseline will determine remaining project tasks and offer new quality checks.

72. You are part of a project that has five teams and serve as team 3's leader. Team 2 has missed several critical deadlines in the past. This has repeatedly caused team 3 to need to crash the critical path. To help resolve this issue, you should meet with:

a. The project manager
b. The project manager and management
c. Team 2's leader
d. The project manager and team 2's leader

Answer: D. If you are having problems significant enough you have had to crash the project (i.e. inject additional resources to meet the deadline), then that implies that you have tried to resolve this problem as best as possible on your end. To prevent this from occurring again and racking up additional costs or jeopardizing other activities on the path, you need to meet with both team 2's leader and the project manager.

73. What is the ideal type of project closure?

    a. Extinction
    b. Addition
    c. Integration
    d. Starvation

Answer: A. Extinction is the ideal type of project closure. Extinction implies a project was successfully completed and accepted. An extinct project does not linger on and has a definitive end point. Answer B, addition, is when a project becomes a permanent feature of the firm, such as evolving into a separate unit within the organization with ongoing operations. Addition necessitates that more resources are added to the project to make it permanent. Answer C, integration, ends one project by merging it with another project, or folding it into an existing work unit. Projects that end in this way see their resources (such as equipment) and team-members either assigned to different projects or returned to their prior roles before the project began. Answer D, starvation, ends a project by withdrawing resources until nothing remains. A decision to starve a project implies failure and an unplanned early end.

74. Project managers can check that project work is indeed complete after a project ends by conducting a _____.

    a. Product verification
    b. Project verification
    c. Procurement verification
    d. Final verification

Answer: A. Product verification is an ongoing process that is only completed when the project is closed. Product verification confirms that project work is indeed complete and matches contract requirements and/or stakeholders' expectations. Answers B, C, and D are not terms used in project management.

75. In Frederick Herzberg's motivation-hygiene theory, hygiene factors deal with:

a. The satisfaction garnered from doing the work and mastering the many skills required in doing that work
b. The environment broadly understood and includes things that prevent dissatisfaction, such as pay, social ties with team-members, and a pleasant work setting
c. That people need to feel they are actualizing their potential to be truly invested in an endeavor
d. Achievement, power, and affiliation as three things team-members need to believe they possess to dedicate themselves to a project

Answer: B. Herzberg's motivation-hygiene theory looks at two motivational theories. The answer, B, refers to hygiene factors, which deal with the environment broadly understood and include things that prevent dissatisfaction, such as pay, social ties with team-members, and a pleasant work setting. The other factors of Herzberg's theory, motivational factors (answer A) deal with the satisfaction garnered from doing the work and mastering the many skills required in doing that work. Answer C refers to Abraham Maslow's hierarchy of needs, while D refers to David McClelland's needs theory.

76. Of the following conflict management methods, which is generally the most time consuming?

a. Problem solving
b. Direct force
c. Withdraw
d. Reconciliation

Answer: A. Problem solving typically takes the longest time, as it requires the project manager to introduce a new set of guidelines to permanently address underlying issues. The belief is there is one correct solution and it must be found. Voices beyond those implicated in the conflict are involved and a solution is constructed from those multiple opinions. The existing conflict is addressed in hopes similar problems in the future will be prevented from a solution derived from consensus. Answer B, direct force, is towards the middle in terms of time spent on conflict management. Answer C, withdraw, is the least time consuming form of conflict management, and answer D, reconciliation, is the second most time consuming method of conflict management, behind problem solving.

77.

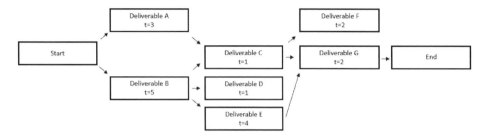

The above chart is known as a:

a. Schedule network analysis
b. Histogram
c. Control chart
d. Network diagram

Answer: D. Combining PERT's emphasis on duration and PDM's emphasis on activity relationships, schedule network analysis is a necessary step to plan out the project schedule. The analysis produces a network diagram (also known as an arrow diagram) in which the duration and dependencies can be viewed.

78. The weather, market, and government regulatory conditions that can affect the firm are all considered:

a. Organizational factors
b. External EEFs
c. Interpersonal factors
d. Internal EEFs

Answer: B. External enterprise environmental factors (EEFs) can make the project more difficult to complete, and include weather, market, and government regulatory conditions that can affect the firm, but not necessarily shape the project directly. Answer A, organizational factors, include the makeup of your firm, its hierarchy, risk management, financial restrictions, safety standards, and rules governing how departments interact with one another. Answer C, interpersonal factors, are qualities of your team members, such as experiences, skills, and cultural sensitivities. Answer D, internal EEFs, can be a firm's culture, history of successfully managing projects, business partners, assets, and knowhow such as subject-matter expertise.

79. A consultancy is most likely a:

a. Non-project based firm
b. Matrix organization
c. Projectized firm
d. Horizontally integrated

Answer: C. Consultancies tend to be project-based or projectized firms. Projectized firms usually have multiple ongoing projects at a time and an existing strategy for tackling projects and a project manager has more authority than in a non-project based firm.

80. _____ is the technique by which high level project requirements are broken down into distinct project activities.

a. Decay
b. Dissolution
c. Decomposition
d. Disintegration

Answer: C. Decomposition is the technique by which high level project requirements are broken down into distinct project activities. Decomposition is a necessary part of making a WBS. Often times, work packages are not obvious in the scope of a project's work. It is necessary for a project manager to decompose project deliverables into manageable work packages and then group similar work packages into activities. Answers A, B, and D are not terms used in project management.

81. Which of the following is an example in which a work authorization system might not need to be used?

a. The project is a very complex
b. The project requires a lot of players
c. The project is small
d. The project requires approval from multiple levels

Answer: C. A work authorization system is a formal, documented procedure that indicates how to authorize work and explains how work should be sequenced at the appropriate time. This is often necessary for complex, large projects that may have a lot of team members and may require high levels to approve work. For small projects, verbal instructions may be enough and a work authorization system is therefore unnecessary.

82. All of these are ways that rolling wave planning shapes work breakdown structure except:

   a. Milestones and targets are further detailed as the project evolves over time
   b. The first WBS has subproject placeholders instead of specific timelines for project completion
   c. It allows teams to work on different phases at the same time
   d. Preventing scope creep

Answer: C. Rolling wave planning is very helpful for project managers because some projects have many subprojects and exact details that cannot be penciled in until a later date. These are dynamic and interdependent tasks. While rolling wave planning enables PMs and their teams to update other phases based on the plans and results of other phases incrementally, it does not actually enable a team to complete work on several phases simultaneously.

83. The difference between earned value (EV) and actual cost (AC) is the formula for:

   a. Schedule variance
   b. Cost variance
   c. Variance at completion
   d. None of the above

Answer: B. The formula provided is for cost variance. If a cost variance is positive, costs are under budget, and if they are negative, then they are over budget.

84. Who is responsible for monitoring a procurement based contract?

   a. Buyers
   b. Sellers
   c. Buyers and Sellers
   d. Lawyers

Answer: C. Since both buyers are sellers have an equal interest in ensuring that both parties live up to contractual obligations, both are equally responsible for monitoring the contract.

85. Which of the following correctly defines risk transfer and risk thresholds?

    a.   A risk transfer occurs when there is a positive effect to a risk occurring, such as monetary savings, and acting in such a way to realize that positive outcome. A risk threshold arises in response to decisions made about a prior risk, such as an exploit strategy.

    b.   Risk transfer occurs when the risk has not occurred and the potential consequences are taken up by a third-party. Risk thresholds define the range within which project leaders are willing to take a chance.

    c.   A risk transfer occurs when there is an opportunity to move the risk outside of the organization. A risk threshold arises in response to prior project experiences.

    d.   Risk transfer occurs when a different project team is put into a riskier project area. Risk thresholds define the range within which project leaders are willing to take a chance.

Answer: B. Only B has the correct definitions. The definitions of exploit strategy and secondary risk are mixed in.

86. A _____ is a program in which projects are grouped together for a common strategic, long-term reason, rather than just efficiency.

    a.   Portfolio
    b.   Hammock
    c.   Milestone
    d.   Summary activity

Answer: A. A portfolio is a program in which projects are grouped together for a common strategic, long-term reason, rather than efficiency. A summary activity (answer D) or hammock (answer B) places many related activities within a shared segment of the project schedule to streamline work. Answer C, a milestone or project phase, is a crucial development in a project. Completing a large deliverable or realizing a deliverable that marks the end of the project are examples.

87. Acquiring the project team is a part of which project management process group?

a. Planning
b. Executing
c. Monitoring and controlling
d. Closing

Answer: B. While it may seem like acquiring the project team is a process that would be in planning, not executing, acquiring your team is actually a part of the executing project management process group. The planning for human resources management occurs during the planning stage; therefore, actually acquiring your team is actually one of the first steps in executing your project.

88. Which of the following is not a technique used in conflict management?
a. Smoothing
b. Direct force
c. Discrete effort
d. Reconciliation

Answer: C. Answer C, discrete effort, is easily measured project work that produces a specific output and is not a part of conflict management. Answers A, B and D are all techniques that can be used in conflict management. Answer A, smoothing, tries to create breathing space in the short-term to see if the issue blows over. Answer B, direct force, doles out a decision to address underlying causes for team friction and demands compliance. Answer D, reconciliation, attempts to have aggrieved parties compromise on something, share the blame, and agree on a solution.

89. _____ planning recommends changes to make in response to a risk that occurs. _____ planning offers strategies to implement in order to make a risk less likely from occurring or to contain the risk's effects if things start to worry the project manager.
a. Contingency, Mitigation
b. Mitigation, Tolerance
c. Contingency, Tolerance
d. Mitigation, Contingency

Answer: A. Contingency planning recommends changes to make in response to a risk that occurs. Contingency plans take the form of: "if this, do this." Mitigation planning offers strategies to implement in order to make a risk less likely from occurring or to contain the risk's effects if things start to worry the project manager. "Tolerance" is not a form of planning.

90. Which of the following is not a need that can result in a project?

    a. Strategic opportunity
    b. Market demand
    c. Lower interest rates
    d. Customer request

Answer: C. There are seven possible needs and demands that can result in a project. They are: market demand, customer request, strategic opportunity/business need, legal requirement, technological advance, social need, and environmental considerations.

91. What is the difference between a project and an operation?

    a. A project is temporary and has a defined beginning and end date, while operations are ongoing and produce the same end result repeatedly
    b. An operation is temporary and has a defined beginning and end date, while projects are ongoing and produce the same end result repeatedly
    c. Projects and operations are interchangeable terms for a series of actions with a defined beginning and end date that produce a product, result or service
    d. Projects and operations are interchangeable terms for a series of ongoing actions that produce a product, result or service

Answer: A. A project is temporary and has a defined beginning and end date, while operations are ongoing and produce the same end result repeatedly. As a project manager you should be dealing with only projects.

92. Iterative life cycles are a good choice for this type of project:

    a. Large projects
    b. Small projects
    c. Simple projects
    d. Projects where deliverables do not need to be delivered incrementally

Answer: A. Iterative or incremental life cycles work best for large or complex projects, or for projects in which deliverables need to be delivered incrementally. If you have a small or simple project, a predictive cycle may be more appropriate, while a project with changing deliverables or time frames may benefit from an adaptive life cycle.

93. A configuration project management plan is most important in which scenario?

a. An ongoing manufacturing process is being repeatedly altered to be in line with multiple managers' efficiency standards. The managers do not try to reconcile their frequently incompatible ideas.

b. A large construction project is rolling out new communication protocols. The protocol involves multiple organizations and contractors.

c. As project monitoring begins for a newly developed software, stakeholders need to be contacted. Many of the stakeholders have not kept up with project progress.

d. During project execution, there are many disagreements about resource sharing. Team-members complain the project schedule did not take their other work into account.

Answer: B. Configuration management is concerned with the rollout of project changes and expanding goals of a project. Answer A describes an ongoing process and so is not a project. Answer C deals with stakeholder involvement. There is no indication that the project is question is large and is being pulled in different directions. Answer D is not very relevant to configuration management since it is an issue within the project team itself.

94. During testing for a newly developed educational curriculum for a school district, a school administrator asks the project manager to swap the order of two sections underneath one of thirty subheadings in the curriculum. The project manager could easily implement this change. What should the project manager do?

a. Examine the effects of this change on other project areas
b. Go through the standardized process of implementing any change
c. Bring the principal's request up to the relevant team members as soon as possible
d. All of the above

Answer: D. No matter how small a requested change is, the standardized process should be followed. This does not mean the change should be ignored. On the contrary, the process in place should expedite consideration of all requests.

95. Which of the following is not a project activity during the project initiation phase?

    a. Designating a project manager
    b. Setting communication requirements
    c. Conducting a feasibility study
    d. Producing a scope statement

Answer: B. Communication requirements are set during the planning phase. The team has to be decided before communication lines and hierarchy can be set. The project cannot begin the planning phase without a designated project manager, so answer A is incorrect.

96. Which of the following is not a category of project selection methods?

    a. Mathematical models
    b. Decision models
    c. Benefit measurement methods
    d. Calculation models

Answer: D. Mathematical models, also known as calculation methods, as well as decision models, also known as benefit measurement methods or benefit analysis techniques, are both two main categories of selection methods. Calculation models is not a reference to project selection methods, although it uses similar phrasing.

97. Which of the following is not a requirements gathering technique?

    a. Brainstorming
    b. Planning sessions
    c. Focus sessions
    d. Focus groups

Answer: C. Brainstorming, planning sessions, and focus groups are all ways of determining potential requirements as a way to develop a cohesive risk management plan.

98. Two techniques used to determine alternative work structure or improve the project in general are:

    a. Life cycle costing and value engineering
    b. Life cycle engineering and value costing
    c. Control costing and estimate engineering
    d. Estimate costing and control engineering

Answer: A. Life cycle costing and value engineering are two techniques that can be used to determine alternatives or improve the project. Life cycle costing looks at a group of costs such as acquisitions, operations, and disposals, for example, to help decide on, or at least compare, alternatives. Value engineering optimizes project performance and cost by primarily looking to eliminate unnecessary costs.

99. What does the term "historical information" refer to in project management?

    a. Archived information on the current project on previous stages of the project
    b. Archived information on a closed project that is archived for reference for future projects
    c. Historical background on aspects such as human capital, funding, or suppliers, for example, that can be used to help inform the current project
    d. None of the above

Answer: B. Historical information is about a project that has been closed and archived, not about the current project. While a lot of the historical information on the project may be background such as information on human capital, funding, or suppliers, it is only information that has been archived as a result of a project's closing.

100. During which project process group is stakeholder influence the lowest?

    a. Planning
    b. Executing
    c. Monitoring and controlling
    d. Closing

Answer: D. Stakeholder influence is greatest during initiating and steadily decreases; therefore, closing has the lowest stakeholder influence. Stakeholders are important to the closing phase, via contributing to lessons learned, but provide the most input on the scope of the project early on.

101. During which project process group should costs be their highest?

a. Planning
b. Executing
c. Monitoring and controlling
d. Closing

Answer: B. During the "executing" group is when costs should be highest. When you consider that this is when the project is actually getting off the ground and the most work is being completed, and when you consider that your staffing levels (often one of your highest costs) need to be their highest when "executing," this makes a lot of sense.

102. At a certain point in project work concerning the total project and one deliverable, the following is true:

- The budget at completion is $2000
- The planned value is $400
- The actual cost is $390
- The earned value is $500

Based on that information, which of the following is correct?

a. The cost variance is -$110
b. The cost performance index is 1.25
c. The schedule variance is -$1600
d. The schedule performance index is 1.25

Answer: D. The question requires knowing and correctly working the relevant equations. For the schedule performance index (SPI), the following work must be done:

$$SPI = \frac{EV}{PV} = \frac{500}{400} = 1.25$$

103. Which of the following is not an input to the define activities process?

    a. Scope baseline
    b. Rolling wave planning
    c. Schedule management plan
    d. Organizational process assets

Answer: B. The scope baseline, schedule management plan, and organizational process assets are all define activities process inputs, as is enterprise environmental factors.

104. When can you start the planning stage?

    a. Once the outline has been approved
    b. When the project steering group has decided full planning can start
    c. Once a planning workshop has been conducted
    d. Once the stage plan has been approved

Answer: B. Although answers A, B, C, and D are all potential parts of the planning stage, even with a completed outline and stage plan a planning stage cannot proceed until the project steering group determines it can. The project steering group can decide to have a delay between the outline and the planning stage.

105. Which of the following descriptions describes an enabling project justification?

    a. The project has benefits that exceed the cost and effort of the project
    b. The project must be run because headquarters requests that it is
    c. The project will allow other operations to produce benefits
    d. The project must be done as a type of maintenance

Answer: C. Even if a project doesn't produce notable benefits on its own, if what it sets in place helps other projects or operations to produce benefits, it is considered "enabling." A describes "benefits," B describes "compliance," and D describes "maintenance" types of project justifications.

106. An example of a project reserve is:

a. Project staff that enter a project only when key project staff are called away for other work
b. Activities that can be reduced in duration if the project is under time pressure
c. Backup materials used if needed equipment is not available
d. Additional sites for the project or parts of the project if the current site is unavailable

Answer: A. An example of reserves is that project staff that enter a project only when key project staff are called away for other work. You should identify who your reserves would be in case of staff members being unavailable, and ensure they will be available to take over a staff member's role at short notice.

107. A project memo should be between a project manager and:

a. the steering group
b. team leaders
c. team members
d. all of the above

Answer: D. A project memo, also known as a project issue, can be between anyone on the project, including team leaders. A project memo can also be used as written communication between the project manager and steering group. Therefore, D is the best answer.

108. What is an authority trigger point?

a. A part of a risk plan in which it is determined when stakeholders must meet to discuss a new or evolving risk
b. A part of a risk plan in which it is determined when a project manager must inform the project steering group of a new or evolving risk
c. A part of the risk plan in which it is determined when a team leader must notify a project manager of a new or evolving risk
d. None of the above

Answer: B. Every risk plan should have authority trigger points, which determine when a project manager needs to notify the steering group of a new or evolving risk instead of simply including a risk in a review or report.

109. In what order should activity processes be performed?

    a. Define activities, sequence activities, estimate activity resources, estimate activity durations
    b. Define activities, estimate activity resources, estimate activity durations, sequence activities
    c. Define activities, estimate activity durations, estimate activity resources, sequence activities
    d. Sequence activities, define activities, estimates activity resources, estimate activity durations

Answer: A. The order of activity processes build on each other: first define activities, then sequence activities, then estimate activity resources, and finally estimate activity durations. Only after these steps have been accomplished can you move on to developing the schedule.

110. Your firm has completed the work on a deliverable as specified by carefully following the procurement statement of work. The product has been formally accepted. After one month, however, your client is displeased with the results. Legally, your firm's contract is:

    a. Incomplete
    b. Complete
    c. Waived
    d. Null and void

Answer: B. Technically, your firm's contact is complete since you followed the procurement statement of work. However, since your client is displeased with the results, your procurement may not be closed and the close procurements process still needs to occur.

111. At what point is risk the highest during a project?

    a. The start of project work
    b. Midway through project work
    c. Close to the project's completion
    d. It depends entirely on the project

Answer: A. The greatest risk is at the start of the project, since so much of the project's phases must build on the success of the previous ones. The beginning of a project also has the greatest potential loss of funding, time, staff and resources. While technically risks can appear or be actualized at any point during the project, these reasons are why the start of the project is so important.

112. Contested project changes can also be known as:

    a. Claims
    b. Appeals
    c. Disputes
    d. All of the above

Answer: D. Contested changes, or changes that cannot be agreed on between parties to a contract, can also be known as claims, appeals or disputes, and can require arbitration if an agreement is not reached.

113. _____ is a forecast of the total cost of the project.
    a. Actual total costs
    b. Estimate at completion
    c. Variance at completion
    d. Budget at completion

Answer: B. Estimate at completion (EAC) is a forecast of the total cost of the project. Answer A, actual total costs (AC), is the exact amount spent at a particular point of the project. Before the project begins, AC is zero. Answer C, variance at completion (VAC), produces the difference between the planned budget and EAC. Answer D, budget at completion (BAC), is the total budget allocated for the project in project plans; the sum of the budget for each project phase.

114. If your stakeholders are very hands-on and plan on actively participating in many aspects of the project, which life cycle category would be most appropriate to implement?

    a. Waterfall
    b. Iterative
    c. Predictive
    d. Agile

Answer: D. If stakeholders plan to be very involved, an agile cycle (also known as change-driven or adaptive method) is the most appropriate life cycle category to consider as it is adaptable to changes in requirements or the environment or other factors.

	SME 1	SME 2	SME 3
Activity 1	RA	I	C
Activity 2	C	R	AI
Activity 3	R	A	C
Activity 4	A	CI	R

115. Which of the following is not true about the chart above?

   a. The chart is a type of RAM
   b. The chart indicates who is notified when an activity is completed
   c. The chart should include the whole project management team
   d. The chart allows a project manager to analyze and control lines of communication

Answer: D. The chart in the question is a RACI chart, which is a type of responsibility assignment matrix (RAM). A RAM ties deliverables or activities to individuals on the project team to provide clear lines of responsibility. RACI charts are a useful and popular type of RAM. RACI is an acronym detailing that chart's usefulness in determining who on the project team is: Responsible, Accountable, Consulted, and Informed through all steps of the project. RACI charts are often quite large and encompass the entire project team, from a project manager's assistant to a subject matter expert (SME) used only once. Answer D is not true of a RACI chart, as analyzing and controlling lines of communication would be done through a basic network diagram and the lines of communication equation.

116. The four benefits of meeting quality requirements are lower costs, higher productivity, increased stakeholder satisfaction and:

   a. Increased customer base
   b. Less rework
   c. Quicker turnaround
   d. Fewer errors

Answer: B. The four benefits of meeting quality requirements are: lower costs, higher productivity, increased stakeholder satisfaction and less rework. Of course, less rework, as well the other benefits, could impact customer interest, turnaround, or the amount of errors.

117. Which of the following is produced when a project's activities are defined?

    a. Activity list, activity attributes, milestone list
    b. Activity list, activity attributes, milestone attributes
    c. Activity list, milestone activities, milestone list
    d. None of the above

Answer: A. The three define activities outputs are: activity list, activity attributes, and milestone list.

118. When deriving activity duration estimates, you might round to the nearest ____, depending on the project.

    a. Hour
    b. Day
    c. Week
    d. All of the above

Answer: D. Which time measurement you will need to round to depends entirely on the project and its complexities.

119. Which of the following is the equivalent of earned value?

    a. The budget at completion multiplied by planned percent of project completed
    b. The budgeted cost of work performed (BCWP)
    c. The actual cost at percent of project actually completed
    d. None of the above

Answer: B. Earned value (EV) or budgeted cost of work performed (BCWP) is what the plan allocated for the actual work completed. The BCWP is another term for earned value. EV determines whether the project is over- or under-budget by calculating what should have been spent in a specific date range. The equation for EV is:

$$EV = BCWP = (BAC)(percent\ of\ project\ actually\ completed)$$

120. Which stakeholder is typically the publisher of the project charter?

  a. Project manager
  b. Project client
  c. Planning group
  d. Executive manager

Answer: B. Typically, the project client is the author or publisher of the project charter. On occasion, the project manager may be asked to author the project charter. However, even if this does occur, the project manager should not be listed as the author or publisher since the charter authorizes you as the project manager. A more appropriate listed author might be an executive manager within your organization.

121. Which of the following is not a common result of risk management?

  a. Recategorizing risks if triggers accumulate
  b. Reserves built into the project's budget are released
  c. A share strategy is executed
  d. Subgroups of the project management team are identified as the risk management team

Answer: D. A risk management team is almost always a separate group from the project team, so it is highly unlikely that a project management team would have a risk management team within it. And the identification is not a result but a preparation. Answers A, B, and C are all common results of risk management.

122. What is the difference between risk appetite and risk tolerance?

    a. Risk appetite details the level of unpredictability project leaders can accept based on risk-reward calculations, while risk tolerance details the level of unpredictability project leaders can accept based on resources
    b. Risk appetite defines the range within which project leaders are willing to take a chance, while risk tolerance details the level of unpredictability that is acceptable based on Monte Carlo numbers
    c. Risk appetite details the level of unpredictability project leaders can accept based on resources, while risk tolerance details the level of unpredictability project leaders can accept based on risk-reward calculations
    d. Risk appetite details the level of unpredictability project leaders can accept based on risk-reward calculations, while risk tolerance defines the range within which project leaders are willing to take a chance

Answer: C. Risk appetite details the level of unpredictability project leaders can accept based on resources, while risk tolerance details the level of unpredictability project leaders can accept based on risk-reward calculations. Answers B and D both have answers that mention risk thresholds, which define the range within which project leaders are willing to take a chance.

123. Project scope summary and the strategic vision behind the project is a part of which of the following?

    a. A portfolio
    b. A work breakdown structure (WBS)
    c. Every project contract
    d. A project statement of work (SOW)

Answer: D. A SOW summarizes at a high level. It should include an analysis of the project scope and project's fit into a firm's strategic goals. The other answer choices include options that are too specific.

124. When is a project considered complete?

    a. When all of the deliverables are completed
    b. When project funds are exhausted
    c. When all stakeholders' needs are addressed
    d. When the project manager decides the project is completed

Answer: C. Even if all deliverables are completed or sponsors or project team members feel the project is completed, stakeholders may ask follow-up questions or have related requests months after the last formal deliverable is completed. Projects have clean end dates but often a stakeholder may ask for some more project work. Contracts often plan for this by requiring firms to answer questions on a product or service well past the formal delivery and transfer of a project.

125. A project manager works for a catering company who is providing catering for a large conference. Although all the food for the conference is prepared and has been delivered to the conference hall, your company can't finalize the food stations' set-up until the project manager coordinating the conference confirms that all other tables and booths are set up. This is an example of which of the following task dependencies?

    a. Finish to start (FS)
    b. Finish to finish (FF)
    c. Start to start (SS)
    d. Start to finish (SF)

Answer: B. This is an example of finish to finish (FF). Although the food may have been already prepared and delivered, your task isn't complete until the other project manager has ensured that the conference hall is fully set up.

126. A project manager is overseeing project in which not all deliverables were able to be mapped out at the start of the project because they are dependent on external factors. As a result, the project manager has subproject placeholders instead of specific timelines for project completion. Therefore, the project manager's work breakdown structure is updated incrementally. This is an example of which of the following planning techniques?

    a. Rolling wave planning
    b. Stage-gate process
    c. Decomposition
    d. Expert judgment

Answer: A. As the name suggests, rolling wave planning entails planning, as the project progresses and details become clearer, in waves. The project manager in this example cannot proceed with updating their work breakdown structure until external factors become clearer.

127. Projects with a net present value (NPV) less than zero should be:

    a. Accepted
    b. Rejected
    c. Paused
    d. Terminated

Answer: B. Projects with an NPV less than zero should be rejected. If it is greater than zero, they should be accepted.

128. What constitutes project completion per project charter rules is deemed:

    a. Acceptance protocol
    b. Acceptance criteria
    c. Handover protocol
    d. Handover criteria

Answer: B. Acceptance criteria details what constitutes project completion per project charter rules and how all parties to the project will express their agreement that the project is complete. Answers A and D are not standard terms used in project management. Answer C, handover protocol, ensures that all necessary project products and artifacts have been successfully signed off on and that documents are archived correctly.

129. A project to revamp a company's onboarding processes is nearly complete. A big, final deliverable is an online tutorial portal. There are weeks of testing planned to ensure the tutorial program meets standards and expectations before delivery occurs. After the testing is completed, the client praises the tutorial as above and beyond their expectations and brings up future opportunities for collaboration.

In the above scenario, the weeks of testing before delivery are considered:

   a. Procurement audit
   b. Product verification
   c. Procurement review
   d. Formal acceptance

Answer: B. Product verification is an internal test, while formal acceptance depends on the project sponsor's satisfaction. Product verification and formal acceptance are similar. Product verification validates that project work is indeed complete and matches contract requirements and/or stakeholders' expectations. The project manager is required to document formal acceptance after delivery and testing. Answers A and C deal with procurements and are unrelated. Procurement review occurs before contract acceptance and is part of the request for proposal phase, evaluating vendors. Procurement audits, meanwhile, measure the completeness of contracts and how they can be improved.

130. Which of the following should not be directly considered when creating a human resource (HR) management plan?

   a. Environmental factors
   b. Organization factors
   c. Personnel policies
   d. Critical success factors

Answer: D. Both enterprise environmental and organization factors should be built into a human resource management plan, another subsidiary of the project management plan. An HR plan details roles and responsibilities of individuals involved in the project and reiterates the firm's policies on reporting requirements, overtime pay, communication protocols, safety requirements, and anything else that governs HR policies. Personnel policies, regarding rules for vacation time, daily breaks, and hiring, firing, and assigning tasks to team-members, should inform the human resource management plan. Critical success factors, however, are typically considered when developing a feasibility study or when creating preliminary project scope statements about the project as a whole.

131. Albert is a project manager at an automobile company. He is in charge of updating a compact car for the upcoming year. He and his team are developing potential changes to the car and are listing these changes' quantifiable and non-quantifiable benefits. Of the benefits listed, which of these is a non-quantifiable benefit?

a. Switching air-conditioning manufacturers will save the automobile company 15% in costs
b. The sleeker redesign of the car's body will attract new customers
c. The compact car will take two hours less total to manufacture
d. The updated audio and screen display will allow the company to charge customers $1,500 more MSRP

Answer: B. The car's redesign is a non-quantifiable benefit because while it may potentially attract new customers and compete with other competitors' updates, the company cannot measure how many more customers are purchasing the car because of its aesthetic changes per the question's facts. A and D are quantifiable benefits that cover financial gain, whereas C is another measurable gain – time saved.

132. When utilizing the milestone technique, a partially complete project activity:

a. Counts towards the percent of project actually completed calculation
b. Counts as halfway towards the percent of project actually completed calculation
c. Does not count towards the percent of project actually completed calculation
d. Counts only the exact amount of the project activity completed towards the percent of project actually completed calculation

Answer: C. The milestone technique measures project progress by the total amount of projective activities that are complete. If a project activity is partially complete, it does not count towards the percent of project actually completed calculation. Answer A is not a technique used in project management, answer B describes the fifty-fifty technique, and answer D describes the percent-complete technique.

133. Consider the formula: $\frac{BAC-EV}{BAC-AC}$. This formula used for:

a. To-complete performance index
b. EAC forecast for work performed at present CPI
c. One-time cost variance EAC
d. Repeated cost variance EAC

Answer: A. The formula in the question is for to-complete performance index (TCPI). The formula for answer B, EAC forecast for work performed at present CPI, is:

$$EAC_{CPI} = \frac{BAC}{CPI}$$

The formula for answer C, one-time cost variance EAC, is:

$$EAC_{CV} = AC + BAC - EV$$

The formula for answer C, repeated cost variance EAC, is:

$$EAC_{RCV} = \frac{BAC}{CPI}$$

134. A project manager compares costs from previous projects to extrapolate costs for a current project. This is an example of:

a. Bottom up estimating
b. Three point estimating
c. Parametric estimating
d. Analogous estimating

Answer: D. Analogous estimating is the least accurate form of estimating. Answer A, bottom-up estimating, is the most accurate and time-consuming form of estimating. However, bottom-up estimates require a lot of information and usually are not able to be done in advance. Answer B, three-point estimating, is more accurate than analogous or parametric estimating. Answer C, parametric estimating, looks at historical data and multiple variables to figure out the cost per unit and then multiples that by the amounts of units needed.

135. Which of the following is not a form of communication that should show up in an issue log?

   a. Upward communication
   b. Lateral communication
   c. Controlling communication
   d. Downward communication

Answer: C. Lateral, upward and downward communication are all forms of communication that should be separated out in your issue log. Answer C, while not a form of communication, sounds similar to the controlling communications plan, which is the total subsidiary plan that structures how a project manager will conduct themselves through each type of message.

136. _____ protects parties against future loss by requiring a reimbursement in the event of cancellation.

   a. Arbitration
   b. Viability decision
   c. Indemnity
   d. Ascertained damages

Answer: C. A prior agreement about an indemnity may obviate the need for arbitration or a trial. An indemnity protects parties to a contract against an unknown future loss by requiring a reimbursement in the event of cancellation. Answer A, arbitration, is when an independent third-party comes in, hears sides of a dispute, and makes a binding decision. Answer B viability decision, refers to a decision made by stakeholders as to whether or not starvation should occur when ending a project. Answer D, ascertained or liquidated damages, is related to indemnity in that it is a set amount, either fixed daily or on some other timelines, to be repaid in the event of a breached contract.

137. A project manager leads a team of 18 people. How many lines of communication exist in this project?

   a. 342
   b. 171
   c. 162
   d. 324

Answer: B. The project manager is part of the team, meaning there are nineteen nodes that need to be plugged into the lines of communication formula:

$$lines\ of\ communication = \frac{18(18-1)}{2} = \frac{19(18)}{2} = \frac{342}{2} = 171$$

138. Critical path method (CPM) charts are most similar to:

    a. GERT charts
    b. Gantt charts
    c. PERT charts
    d. RACI charts

Answer: C. A PERT chart is almost identical to a CPM chart. However, CPM charts are more complex than PERT charts because a CPM chart breaks down the earliest possible times for a task's completion, while a PERT chart only displays the timing of each step of the project and their sequence.

139. The chairperson of a project steering group is the:

    a. Team leader
    b. Project manager
    c. Project user
    d. Project sponsor

Answer: D. The chairperson should always be the project sponsor. Project sponsors can also be known as the project director, project executive or simply executive, or a senior responsible owner (SRO).

140. _____ is project work that is understood to be continuous and not divided into discrete work packages. By contrast, _____ is easily measured project work that produces a specific output. _____ is project work measured in time and is not necessarily tied to a specific deliverable.

    a. Discrete effort, apportioned effort, level of effort
    b. Apportioned effort, level of effort, discrete effort
    c. Level of effort, discrete effort, apportioned effort
    d. Apportioned effort, discrete effort, level of effort

Answer: D. Apportioned effort is project work that is understood to be continuous and not divided into discrete work packages. By contrast, discrete effort is easily measured project work that produces a specific output. Level of effort is used to determine how long an activity will take and depends on the skillset of the team-member.

141. Which of the following is a shared property of project closure and project completion?

    a. They are both continuous processes and do not have an exact date
    b. Depending on the project, they occur at the same time
    c. They both depend on the project's stakeholders
    d. All of the above

Answer: C. Project completion is not the same as project closure. To eliminate these options, you must know the differences. After the final delivery stage is carried out, all project activities are complete. But between project completion and the closure date, all project work must be evaluated so lessons are learned, and future projects can go smoother. Only when stakeholders accept or deny project deliverables is a project complete; and for a project to be closed, stakeholder feedback needs to be integrated. Both completion and closure happen at an exact time when they are finished.

142. Handoffs can also be known as:

    a. Technical transfers
    b. Feasibility changes
    c. Phase sequences
    d. None of the above

Answer: A. Handoffs are also known as technical transfers. While they occur as a result of the changing of phase sequences, they are not called phase sequences.

143. Your business case can be produced by considering the project's:

    a. Justification
    b. Benefits
    c. Roles
    d. Scope

Answer: A. In order to produce the project's business case, you must determine if and how the project is justified. Simply wanting to do the project is not justification in and of itself. While B, benefits, is relevant to the business case in that benefits are set down in the business case, the correct answer is A. The business case, and therefore benefits, cannot be determined without the project's justification being laid out first.

144. Which of the following is a common property of a project's procurement process?

a. Requests to perform
b. Invitations to auction
c. Requests for quotation
d. Invitations to inform

Answer: C. Requests for proposal (RFP), requests for information (RFI), invitations to bid (IFB), and requests for quotation (RFQ) are typical ways an organization finds a contractor from which to procure desired services. The other answer choices jumbled the acronyms.

145. Which of the following is not a project activity during the project planning phase?

a. Norming project work
b. Identifying risks
c. Analyzing assumptions
d. Creating a Responsibility Assignment Matrix

Answer: A. Norming project work requires project work to be ongoing, and so is a property of the execution phase. All the other options are part of the planning phase.

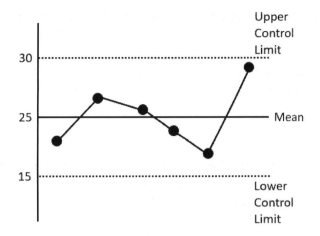

146. For the type of chart above, one standard deviation is ___ above and below the mean and two standard deviations is everything within ___ of the same mean.

    a. 34%, 47.2%
    b. 20%, 40%
    c. 28%, 42.5%
    d. 40%, 60%

Answer: A. For the question's chart, a control chart, one should know that one standard deviation includes everything 34% above and below the mean (68% total), while two standard deviations includes everything within 47.2% of the same mean (95% total).

147. Which of the following is not a project activity beginning during the project execution phase?

    a. Relying on checklists, the Delphi technique, and other strategies to handle risks
    b. Norming project work
    c. Under taking multi-criteria decision analysis
    d. The project manager establishing a consistent leadership style

Answer: D. The project manager must establish a consistent leadership style, whether laissez-faire or theory x, during the planning phase. The project manager will first deal with many team members during planning and even initiation. These earlier contacts are when impressions are formed. The next best answer, C, involves the onboarding of new project team members during execution.

148. Which of the following is not a project activity beginning during the project controlling and monitoring phase?

a. Performance appraisals
b. Rebaselining
c. Variable sampling
d. Relying on integrated project management software

Answer: D. Integrated project management software regulates all project work and is not begun during controlling and monitoring. All of the others are controlling or monitoring efforts that regulate project execution.

149. Which of the following is not a project activity during the project closing phase?

a. Product verification
b. Producing project closure plans
c. Procurement closure
d. Writing a lessons learned report

Answer: B. Like all plans, project closure plans are made during the planning phase before project work begins. Those plans may be modified in the closing phase, but are not produced during closing.

150. Who are the two groups with whom a project manager should prioritize consistent communication?

a. Project drivers and key stakeholders
b. Project drivers and contractors
c. Key stakeholders and functional managers
d. Contractors and functional managers

Answer: A. Project drivers and key stakeholders are the two most important groups that a project manager should prioritize consistent communication. Project drivers are the people most listed as "responsible" or "accountable" on a variety of tasks. In a broader sense, stakeholders are any individuals with an interest in the project's outcome. Contractors and functional managers can fall under either of these umbrellas, which is why A is correct.

151. If your organization already has a business in mind for procurement and skips or expedites the process of requesting and reviewing, this is called:

a. Prerequisite
b. Preassignment
c. Prebidding
d. Prerequesting

Answer: B. Preassignment skips or at least expedites the process of requesting and reviewing. Often, a buyer will have a partner organization in mind and offer them the entire contract or part of it first. When this occurs, specific individuals are often requested with whom the buyer had a positive prior experience.

152. A process by which a project manager lets the project team know what metrics they will be checking to measure progress the next time the project is reviewed is called a:

a. Performance appraisal
b. Progress review
c. Forward view
d. Intermediate step

Answer: C. A forward view is a process by which a project manager lets the project team know what metrics they will be checking to measure progress the next time the project is reviewed. If the project manager meets with team members the first week of each month, the next month will be discussed along with the past month during reviews. Answer A, performance appraisals, are informed by forward view. Performance appraisals encompass the many forms reviews of project work can take, including formal and informal methods. Answer B, progress review, is a made up term, and D, intermediate step, which are not major milestones but day-to-day activities.

153. The project steering group is made up of:

a. Sponsor, project users, and project suppliers
b. Sponsor, project users, and project administration
c. Project users, project suppliers, and project managers
d. Project managers, team leaders, and project administration

Answer: A. The project steering group is made up of the sponsor, project users, and project suppliers.

154. What is the primary role of a project expeditor?

    a. Support staff and coordinate communication
    b. Determine and delegate ideal tasks to different team members
    c. Authorize the project and provide guidance to the team
    d. Facilitate interaction and productive communication

Answer: A. A project expeditor is a supportive role to the project manager and the team. Their primary duty is to help coordinate communication inside and outside the team and support the team as needed. Answer B is incorrect because a project expeditor cannot make any decisions on their own authority. Answer C refers to the role of the project sponsor, while answer D refers to the primary role of the project manager.

155. Which of these inputs is not a subsidiary plan?

    a. Scope management plan
    b. Risk management plan
    c. Activity list
    d. Procurement plan

Answer: C. All of the inputs that make up the larger project management plan are known as subsidiary plans. Subsidiary plans include: the scope management plan, risk management plan, communication plan, procurement plan, and reserve analysis plan, for example. The activity list covers the scope of work for each scheduled activity.

156. Float is also known as:

    a. Buffer
    b. Leveling
    c. Drift
    d. None of the above

Answer: D. Float is also known as free slack. Float refers to the latest possible delay a deliverable can be started without delaying the project.

157. Analogous estimating is:

a. A form of top-down estimating
b. A form of bottom-up estimating
c. A form of parametric estimating
d. It depends on the project

Answer: A. Analogous estimating is a form of top-down estimating. Analogous estimating examines the budgets of similar completed projects within the firms or accessible budgets (for example, government contractors) from outside the firm. Answer B is incorrect since it is the opposite of top-down estimating (bottom-up estimating). Answer C, parametric estimating, is another form of top-down estimating. Answer D is incorrect since analogous estimating is always top-down estimating.

158. The total amount needed to sustain all of a project team's resources, from electric bills to every salary is called:

a. Sunk cost
b. Budget baseline
c. Full burdened rate
d. Budgeted cost of work scheduled

Answer: C. The fully burdened rate is the total amount needed to sustain all of a project team's resources, from electric bills to every salary. Answer A, sunk cost, is an amount which has already been spent, is not affected by estimates and not accounted for in control thresholds. A sunk cost can also be money spent to purchase capital for a past project that will again be used in a new project. Answer B, budget baseline, is the original, final cost estimate. Answer D, budgeted cost of work scheduled (BCWS) provides totals for all types of specialized work over the entire scope of a project.

159. Which of the following can be used to attain procurements?

a. Invitation to bid
b. Requests for information
c. Requests for proposal
d. All of the above

Answer: D. Procurements can be attained from potential suppliers by issuing requests for proposal (RFP), requests for information (RFI), and invitations to bid (IFB). Requests for quotation (RFQ), not a listed answer, can also be used to obtain procurements.

160. Your firm has been chosen by a new client because the client did not see the profit estimate they anticipated from a prior project conducted by a different firm. The client feels that the previous firm did not do enough to ensure that the product would be profitable. Which of the following should your firm conduct to ensure the client's needs are compatible with the firm's goals before taking on the client?

a. Capital appropriations plan
b. Legal contract analysis
c. Cost-benefit analysis
d. Feasibility statement

Answer: D. While all of the possible choices should be conducted when taking on a project, only answer D directly ensures that a client outlines, at a granular level, what they hope the project to accomplish – including profitability. Answer A, capital appropriations plan, simply itemizes the firm's budget expenditures, which is not directly connected to a product's profitability. Answer C, a cost-benefit analysis, can complement a feasibility study, but only covers the risks the firm can have if taking on the project, not whether or not the product can be more profitable. While answer B, a legal contract, specifies the obligations your firm has to the client, it can only help enforce what the feasibility statement defines.

161. A work flow diagram shows deliverables in the order:

a. Of importance
b. In which they must be completed
c. In the order of which they must be started
d. It depends on the type of work flow diagram

Answer: B. A work flow diagram shows deliverables in the order in which they must be produced.

162. While configuration management coordinates the rollout of approved changes to the activities and goals of the project, more substantial changes to project plans, or whenever a new project plan is adopted, needs _____ to determine remaining project tasks and offer new quality checks.

    a. Critical path change
    b. Rebaselining
    c. Forward projections
    d. Control processes

Answer: B. For more substantial changes to project plans, or whenever a new project plan is adopted, rebaselining is necessary. The new baseline will determine remaining project tasks and offer new quality checks. Any time a change occurs to critical path project activities, a new baseline will be needed as critical path changes will, by definition, delay the entire project. Answer A, critical path change, is not strictly a project management term. Answer C, forward projections, are used many times over the course of a project, but are essential when controlling project execution and when baselining is necessary. Answer D, control processes, is too vague, although configuration management is a type of control process, for example.

163. In conflict management, what is the difference between smoothing and withdrawing?

    a.   Smoothing involves a project manager stepping in to "smooth" out issues between team members by introducing a new set of guidelines to permanently address underlying issues, while withdrawing involves a project manager creating breathing space between the two parties in the short-term to see if the issue blows over.

    b.   Smoothing involves a project manager creating breathing space between the two parties in the short-term to see if the issue blows over, while withdrawing implies that the project manager ignores rather than addresses the conflict.

    c.   Smoothing involves a project manager creating breathing space between the two parties in the short-term to see if the issue blows over, while withdrawing involves a project manager taking a step back or "withdrawing" from the conflict to allow team members to develop a new set of guidelines to permanently address underlying issue.

    d.   Smoothing involves a project manager ignoring rather than addressing the conflict, while withdrawing involves the project manager creating breathing space between the two parties in the short-term to see if the issue blows over.

Answer: B. In conflict management, withdrawing implies a conflict is ignored rather than addressed. This could occur when one party refuses to discuss the conflict and merely adjusts, a party leaves the project, or it is decided the conflict was a minor issue that does not warrant attention. Smoothing tries to create breathing space in the short-term to see if the issue blows over. It minimizes problems and tries to reinforce ground-rules in communication that may have lapsed. The other option presented in this question refers to problem solving. Problem solving requires the project manager to introduce a new set of guidelines to permanently address underlying issues. The problem-solving belief is there is one correct solution and it must be found.

164. Which of the following is an internal dependency?

    a.   An event that occurs outside the project scope that triggers a schedule change
    b.   A sequence of tasks not dictated by the nature of the work, but rather agreed to within the project team
    c.   An event that occurs within the project team, but not tied to the execution of tasks, that can change the project schedule
    d.   Tasks that have to be done in an exact order

Answer: C. That is the correct definition of an internal dependency. An example of a delay to an internal dependency might be a project team member having to leave their job for family reasons, resulting in postponed work for the tasks they handled. It occurred within the project team but is still not directly related to the project work. Answer A is an external dependency, answer B is a discretional dependency, and answer D is a mandatory dependency.

165. Team involvement is most essential in:

    a.   Work breakdown structure (WBS) creation
    b.   Reserve analysis
    c.   Making a viability decision
    d.   Scope validation

Answer: A. A WBS provides a way for team members to understand how their work fits into the whole. Involving team members with specialized skills will also allow work to be broken down in the most efficient manner. All of the other options need authority greater than team-members'.

166. An automotive industry project is under-budget and ahead of schedule. Team members' work efficiency is increasing. Project stakeholders are content and their feedback is positive. Project stakeholders are so satisfied that they are offering ideas for additional but modest features to add to the project. The project manager obliges and unilaterally organizes work to realize stakeholders' suggestions.

The above scenario is an example of:

    a.   Model project management
    b.   Control scope
    c.   Efficient communication
    d.   Scope creep

Answer: D. Scope creep comprises all unauthorized changes to the project plan that are not authorized as they happen. A project manager unilaterally (on their own) adding capabilities is a clear example. It is not clear if efficient communication is occurring. Efficient communication stresses the timeliness, in addition to the relevancy, of communication.

167. If the schedule performance index (SPI) is 1.1 and the cost performance index (CPI) is 1, which of the following must be true?

   a. The project is behind schedule
   b. The project is right on schedule
   c. Planned value is greater than actual cost
   d. Planned value is less than actual cost

Answer: D. Both SPI and CPI are quotients with earned value (EV) as the numerator:

$$SPI = \frac{EV}{PV} \quad \& \quad CPI = \frac{EV}{AC}$$

Since the SPI's quotient is greater than CPI's, PV (planned value) must be greater than the opposite denominator (actual cost). An SPI of 1.1 indicates that the project is ahead of schedule. A CPI of is exactly 1 indicates that the project's planned budget is being followed to the penny.

168. Network diagrams are meant to display:

   a. Critical paths
   b. Dependencies
   c. Relationships in-between tasks
   d. All of the above

Answer: D. Networks diagrams show the interconnectedness of project work. All of the options speak to that purpose.

169. The amount of time an activity can be delayed without extending project timelines is:

a. An activity's float
b. A critical path
c. Maximax criterion
d. Resource crashing

Answer: A. Float refers to the latest possible delay a deliverable can be started without delaying the project. The float of an activity is severely restricted if the activity is on a critical path. A delay on the critical path will delay the entire project. Resource crashing adds additional resources to expedite the project's critical path. Maximax criterion has to do with risk, emphasizing the possibilities inherent in a risk.

170. Towards the middle of project work, the client who commissioned the project doubles in size due to a previous investment paying off. The larger firm now has much more money to spend on the project. Instead of just continuing the project, however, the client offers to buy the performing organization and make the project work a permanent part of their business. The performing organization rejected the bid. In a second offer, the client asked the project to be merged with another two projects. The performing organization accepted and merged projects, thereby ending the original project. The larger project was successfully and fully completed.

Which of the following correctly describes the scenario above?

a. After an integration proposal was rejected, an addition offer was accepted
b. After an addition proposal was rejected, an integration offer was accepted
c. After one addition proposal was rejected, another addition offer was accepted
d. After one integration proposal was rejected, an integration offer was accepted

Answer: B. Addition is when a project becomes a permanent feature of the firm, such as evolving into a separate unit within the organization with ongoing operations. That was the first offer. Integration ends or modifies one project by merging it with another project, or folding it into an existing work unit. Projects that end in this way see their resources (such as equipment) and team-members either assigned to different projects or returned to their prior roles before the project began.

171. A project manager is facing a make-or-break point in project work. After a number of close calls, progress has slowed and the next set of tasks are crucial. If the next set of tasks cannot be completed on time, the project may be starved. All of these imminent tasks are dependent on one another such that one activity can only be completed when another activity is also completed. In a precedence diagram, these tasks have which type of relationship?

a. Finish to start (FS)
b. Finish to finish (FF)
c. Start to start (SS)
d. Start to finish (SF)

Answer: B. Tasks that are dependent on one another to finish at the same time are in a finish-to-finish relationship. Finish to start describes a relationship where one activity cannot start until another one finishes. Start to Start describes a relationship where one activity cannot start until another one starts. Start to finish describes a relationship where one activity cannot finish until another one starts.

172. Which of the following is an example of a fixed cost?

a. Project startup costs
b. Reserve funds
c. Administrative costs
d. Bottom-up costs

Answer: A. Fixed costs do not change as the project matures, grows, or gets smaller. These changing costs are labeled variable costs. Startup costs have to do with the implementation of any project plan—getting it off the ground. Reserve funds change during the project, depending on how well plans pan out. Administrative costs concern paying out project team salary, which can change depending on hiring. There are no bottom-up costs; there is bottom-up estimating.

173. What is a work breakdown structure?

    a. The work breakdown structure is a framework for understanding how part of a project affects the whole
    b. The work breakdown structure is the total amount needed to sustain all of a project team's resources, from electric bills to every salary
    c. The work breakdown structure is a chart that ties deliverables or activities to individuals on the project team to provide clear lines of responsibilities
    d. The work breakdown structure is a deliverable based breakdown of a project into lesser components

Answer: D. A work breakdown structure is a deliverable based breakdown of a project into lesser components. The work breakdown structure defines the scope of the project, is focused on deliverables, shows how deliverables are related to one another, and is the basis for planning after a project is initiated. A WBS should include four levels: the project, deliverables, work packages, and activities or components. The latter two are considered level-three while deliverables, phases, and sub-projects are considered level-two. Level-one is the project.

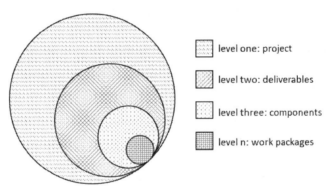

level one: project

level two: deliverables

level three: components

level n: work packages

174. Which of the following is NOT a major part of project life cycles?

    a. Initiation
    b. Execution
    c. Milestone
    d. Closure

Answer: C. A project's life cycle phases are initiation, planning, execution, monitoring and controlling, and closure, whereas a milestone is a crucial development in a project.

175. Which of the following theories referring to cost of quality (COQ) focuses on output and believes an optimum number of allowable defects rounds out to 3.4 per 1,000,000?

    a. The plan-do-study-act (PDSA) cycle
    b. Total quality management (TQM)
    c. The Deming cycle
    d. The six sigma (6σ) measurements

Answer: D. Six sigma focuses on output and believes an optimum number of allowable defects rounds out to 3.4 per 1,000,000. COQ refers to the expenditure required to achieve a specific standard in the project. Estimates for varying quality are usual integral parts of three-point estimate's best to worse case cost scenarios. It is mostly used in projects with a physical output.

176. You are a project manager for a product that was due in the third fiscal quarter. However, your client wants the project to be delivered in the second quarter. The client is willing to invest in more resources to ensure the project is completed sooner. Which of the following strategies should you employ first to get the entire project delivered at a quicker pace?

    a. Reverse resource allocation
    b. Resource crashing
    c. Resource leveling
    d. Fast tracking

Answer: B. Although fast tracking would typically be a first step before resource crashing, because the client has indicated they are willing to invest more cash into the project, resource crashing is a better option. Fast tracking has increased, but resource crashing allows fast tracking and many other strategies to expedite production.

177. What is the difference between total float and free float?

    a. Total float refers to the amounts of time an activity can be postponed without delaying the start of another separate activity, while free float refers to the amount of time an activity can be postponed and begun later without pushing back the overall project end date

    b. Total float refers to the amount of time an activity can be postponed and begun later without pushing back the overall project end date, while free float refers to the amounts of time an activity can be postponed without delaying the start of another separate activity

    c. Total float modifies a relationship between project activities and accelerates work on a dependent successor task when the preceding task is bogged down, while free float extends the time between two dependent tasks

    d. Total float extends the time between two dependent tasks, while free float modifies a relationship between project activities and accelerates work on a dependent successor task when the preceding task is bogged down

Answer: B. Total float refers to the amount of time an activity can be postponed and begun later without pushing back the overall project end date, while free float refers to the amount of time an activity can be postponed without delaying the start of another separate activity. Answers C and D refer to leads and lags. A lead modifies a relationship between project activities and accelerates work on a dependent successor task when the preceding task is bogged down. A lag extends the time between two dependent tasks. A lag is necessary when two preceding tasks feed into one successor task and one of these preceding tasks is completed much earlier than the other.

178. What is another term for soft logic?

    a. External dependency
    b. Discretionary dependency
    c. Mandatory dependency
    d. Internal dependency

Answer: B. Discretionary dependency, also known as soft logic, allows the project manager to set an order between two tasks because there is not a mandatory dependency between those two tasks. Mandatory dependency is also referred to as hard logic, which sets a required order between two tasks because there is a mandatory dependency between the two, such as a contract.

179. A project manager's budget plan does not account for the varying exchange rate between the United States' dollar and the United Kingdom's pound, where the product will be produced. The project manager did not adequately consider which of the following concepts when budget planning?

    a. Opportunity cost
    b. Economic profit
    c. Level of accuracy
    d. Cultural sensitivities

Answer: C. While it is possible that this error will affect the initial opportunity cost or ultimate economic or accounting profit of the project, the main blunder this project manager had was a lack of accuracy. Maintaining a consistent level of accuracy (similar rounding or same metrics, for example) is essential for a robust budget plan.

180. Which of the following is least important when developing a procurement plan?

    a. Vendor processes
    b. Timing
    c. Roles and responsibilities
    d. Lines of communication

Answer: D. The procurement plan is a subsidiary plan that decides what resources will be bought, when they will be bought, and the vendors. It affects teamwork, budget, and schedule planning. While understanding the organization's lines of communication within the totality of individuals working on a project is a best practice, it is not directly connected to the procurement plan.

181. The majority of a project manager's time is taken up with:

    a. Planning
    b. Communicating
    c. Budgeting
    d. Quality control

Answer: B. The majority of a project manager's time, up to ninety percent in some studies, is taken up by communication. While planning (answer A), budgeting (answer C) and quality control (answer D) can all be time-consuming, communication is essential to each of these in ensuring that everything runs smoothly and that all parties are well-informed about the progress of the project.

182. A _____ is the most detailed day-to-day description of project work.

    a. Project activity
    b. Project life cycle
    c. Work breakdown structure
    d. Work package

Answer: D. A work package is the most detailed day-to-day description of project work. Answer A, project activity, can be any one of the three lower levels of a project. An activity is simply a specific portion of the project work that has separate parts that are similar. Answer B, project life cycle, is the sum of each project activity, from start to finish. Answer C, work breakdown structure (WBS), defines the scope of the project, focuses on deliverables, shows how deliverables are related to one another, and is the basis for planning after a project is initiated.

183. Which of the following is a characteristic of adequate project scoping?

    a. A traceability matrix is developed to allow understanding of how deadlines were decided
    b. Project assumptions are hypothesized before the project develops
    c. Decomposition of project activities breaks down to better anticipate workloads
    d. The sunk cost should be determined before a legal contract is drawn up

Answer: C. The project scope details all the work that must be done to meet a project's entire deliverables. Answers B and D refer to project assumptions and sunk cost, respectively, which can only be determined once a project is being executed. Answer A refers to a traceability matrix, which is typically used to assess if the current project requirements are being met.

184. A summary activity is also known as a:

a. Hammock
b. Portfolio
c. Milestone
d. None of the above

Answer: A. A summary activity is also known as a hammock. A summary activity or hammock places many related activities occurring between two dates together to streamline work. Answer B, a portfolio, is a program in which projects are grouped together for a common strategic, long-term reason rather than efficiency. Answer C, a milestone, is a crucial development in a project. Completing a large deliverable or realizing a deliverable that marks the end of the project are examples. Identifying project milestones will facilitate effective time-management. Milestones have specific start and end dates. Milestones can also be referred to as a project phase.

185. A project manager at a project-based firm is deciding whether to recommend a project to redesign a product for a prospective client. The net present value yields a negative value but the president of the firm and many stakeholders think the prospective client could introduce them to similar clientele if the product redesign is successful. What should the project manager do?

a. Recommend against the project for the time being
b. Recommend for the project
c. Present the NPV alongside a separate report on the value of the product's redesign, hear feedback, and then decide on a recommendation
d. Let the board or any other entity decide whether to proceed or not

Answer: A. The NPV (net present value) includes profit and *value*. If an NPV is negative, a project should not be recommended. Ascertaining the value of a project, the provision in answer C, must be a part included in an NPV. In the initiation stage, a project manager is responsible for recommending a decision. The firm may go in a different direction regardless.

186. Project managers use earned value measurement to:

a. Remain current with budgetary and spending plans
b. Get objective information on project work, free from functional managers' biases
c. Get up-to-date reports on project performance
d. All of the above

Answer: D. Earned value measurement (or management) is a dexterous tool. The three features highlighted here are the main uses.

187. A large, multi-year marketing project is subject to quarterly audits. These audits are supposed to be all-encompassing but usually focus on just the financial aspects of the project. Only two weeks after the most previous audit, the project manager and team find out that an extensive quality-audit is to be undertaken. There are multiple complaints amongst the project team and opposition to this unannounced quality audit. Why is it important to cooperate fully with this new audit?

a. Achieving a high level of quality is the overriding purpose of a project
b. Having an outside source look for efficiency and effectiveness shortcomings will streamline work and benefit all parties
c. The project team is to execute not oppose project demands
d. A quality-audit can ensure the project is complying with industrial standards

Answer: B. Quality audits are a tremendous boost to project efficiency and effectiveness. In this case, it is the unexpected nature that is a nuisance but it is still a positive development. For example, finding a small inefficiency will save the executing business unneeded spending. That saved time can be spent producing a better result. The other choices are flawed. A project has multiple purposes for the parties that are attached to it. "Quality" does not capture those purposes. Project teams can oppose policies and provide valuable input to correct a flawed demand. Industrial standards are not necessarily pertinent to quality audits and the specifics of the question scenario.

188. Which of the following correctly identifies the key differences between a histograms and Ishikawa Diagrams?

    a. Histograms are used to display occurrences of an issue, while Ishikawa Diagrams try to find the origins of project problems
    b. Histograms are used to show the history of an issue, while Ishikawa Diagrams try to find the origins of project problems
    c. Histograms are used to display occurrences of an issue, while Ishikawa Diagrams try to find problems before they occur
    d. Histograms are used to display occurrences of an issue, while Ishikawa Diagrams try to find the cost of project problems

Answer: A. Histograms are bar graphs in which the horizontal axis is an occurrence category and the vertical axis is that category's frequency. A cause and effect (or a fishbone or Ishikawa) diagram's purpose is to find the origins of project problems. Two examples are below: histogram first, and Ishikawa diagram below it.

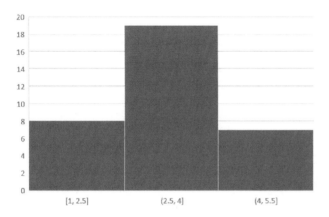

189. A critical part of a lessons learned report is the _____, which analyzes the strengths and weaknesses of how changes to the project plans were handled in the project.

    a. Final budget
    b. Schedule performance
    c. Project logs
    d. Controls review

Answer: D. A critical part of a lessons learned report is the controls review portion. The controls review analyzes the strengths and weaknesses of how changes to the project plans were handled in the project. It also proposes ways to improve project controls. Answers A, B, and C all include information that is important to include in the lessons learned report.

190. When identifying stakeholder inputs, which of the following is an environmental factor you may want to pay notice to?

    a. Organizational structure
    b. Company culture
    c. Governmental/industry standards
    d. All of the above

Answer: D. The organizational structure, company culture, and governmental or industry standards are all important environmental factors worth paying close attention to.

191. Instead of traditional written reports, team leaders or project managers can use digital _____, such as work flow diagrams, to show delivery.

    a. Gauge reporting
    b. Presentations
    c. Dashboard reporting
    d. Demos

Answer: C. While answers B and D are both potential forms of reporting other than actual written reports, work flow diagrams are a form of dashboard reporting. Answer A, gauge reporting, is not an actual term used in project management (although dashboard reporting can use "gauges" to show stages). Therefore, the only correct answer is C.

192. As a project manager, you need to help your company decide between taking two projects. Both projects will take place during a period of six months and both will generate a cash inflow of $75,000. However, Project A requires 50 hours of billable work per week, while Project B requires 40 hours of billable work per week. Which project would you advise stakeholders to take, ceteris paribus?

    a.   Project A
    b.   Project B
    c.   They both have the same cash inflow per year so both are equally advisable
    d.   Not enough information to make a decision

Answer: B. Ceteris paribus means "other things equal" in Latin. Therefore, you only have the information available. Mathematically, the inflow is equal. However, best practices for a project manager to go for the option that requires fewer hours for the team since that is more money per hour.

193. You are taking over from a project manager moved to another department early on in the initiation phase. The departing project manager tells you that she was working on the project's objectives, the desired deliverables, and anticipated timelines before the firm commits to the project. The project manager was working on which of the following?

    a.   The subsidiary plan
    b.   The project management plan
    c.   The project charter
    d.   The preliminary project scope statement

Answer: D. The preliminary project scope statement describes a project's objectives, the desired deliverables, and anticipated timelines before a firm commits to a project.

194. A linear responsibility chart:

   a. Is mostly drawn up for industrial or manufacturing projects in which minor physical defects could set the project back.
   b. Displays the relationships between project activities.
   c. Ties deliverables or activities to individuals on the project team to provide clear lines of responsibilities.
   d. Has a vertical axis representing time and horizontal bars reflecting how long a project task is projected to take.

Answer: C. A linear responsibility chart (LRC), is a type of responsibility assignment matrix (RAM) which ties deliverables or activities to individuals on the project team to provide clear lines of responsibilities. Answer A refers to a control chart. Answer B refers to a precedence diagram, and answer D refers to a Gantt chart.

195. Which of the following is NOT an input that would be relevant to estimating costs?

   a. Human resource management plan
   b. Scope baseline
   c. Core competencies
   d. Risk register

Answer: C. Answers A, B, and D are directly relevant to estimating costs. Core competencies, however, differentiate one business from another. A team's education and experience determines core competencies. A new project will either strengthen a business's existing core competencies or create new ones.

196. Rolling wave planning shapes the work breakdown structure (WBS) in what way?

   a. The first WBS has subproject placeholders instead of specific timelines for project completion
   b. Milestones and targets aren't detailed until the project evolves over time
   c. Some deliverables cannot be mapped out at the start of a project, so the WBS is not completed at a single point
   d. All of the above

Answer: D. Some projects have many subprojects and exact details that cannot be penciled in until a later date. These are dynamic and interdependent tasks. Therefore, the planning occurs in waves and the WBS is updated incrementally.

197. One way Monte Carlo analysis can be used is to:

a. Indicate resource needs
b. Assess project risk probabilities
c. Address quality
d. Indicate which team-member is least efficient

Answer: B. Monte Carlo analysis is used to assess project risk as it feeds into Graphical Evaluation and Review Technique. Monte Carlo analysis does not address quality or resource needs, nor staff needs. While a Monte Carlo analysis can help assess whether an estimate for an activity needs to be revised, it does not actually indicate what an activity estimate ought to be.

198. A project is being considered that requires the involvement of a social media marketer who frequently makes spelling and grammar errors and is sensitive to criticism. During the initiation stage, how should a project manager expect to improve this social media marketer's performance?

a. Plan on either decreasing or increasing the social media marketer's future pay based on performance
b. Clearly state the social media marketer's duties ahead of time
c. Get the social media marketer's functional manager involved early on and ensure the manager understands the scope of the project
d. Get the social media marketer's functional manager involved early on and ensure the manager evaluates the social media marketer's performance based on project performance

Answer: D. A project manager (in that one role) cannot reward or determine pay for team-members, so answer A is incorrect. Even without work difficulties, answers B and C should be done. Following a difficult relationship, a project manager should be proactive and ensure managers tie incentives and disincentives to project performance.

199. During an information technology project, problems arise about quality and grade. During a meeting, a subject-matter expert asks: "How are grade and quality different?" How would you answer?

    a.   Grade measures condition, while quality measures features
    b.   Grade measures features, while quality measures condition
    c.   Grade and quality measure the same things but in different ways
    d.   Grade and quality are synonyms

Answer: B. Quality has to do with condition (such as durability). Grade has to do with features, which are a choice. A project cannot begin with equipment with an incorrect grade. Lower-quality equipment would make a product with fewer features, not necessarily lower-quality.

200. A client that your firm has a positive history with praises your firm's handle on project processes and avoiding setbacks. They ask for you to eliminate project management costs since they inflate costs "unnecessarily" from their point of view. If the answers below are all possible or true, which of the following is the best way to respond to this client?

    a.   Agree with your client and remove project management costs
    b.   Remove some costs, such as those associated with communications and meetings
    c.   Remove all costs but the project manager's salary
    d.   Explain the costs sustained on previous projects that did not use project management

Answer: D. Informing your client about the costs that past projects without project management processes sustained will help your client understand the risks, and potentially higher costs, of not using all project management processes during a project.

<u>Notes</u>

<u>Notes</u>

Notes

Thank you

Thank you for purchasing *PMBOK ® Guide and PMP Exam Prep Book 2018-2019: Study Guide on the Project Management Body of Knowledge with Practice Test Questions for the Project Management Professional Exam by Robert P. Nathan*. It is my hope that this book has assisted you well on your journey to becoming a certified PMP.

This book is my project. I do not consider it closed until you are satisfied. Unlike other books written by teams of corporate writers, I am looking for ways to continually improve my work and approach to project management. Let me know if there is something I can help you with. Email your questions and comments to: robert@pmppeternathan.com.

Additionally, if you found this book helpful, please leave a positive review on Amazon. In an industry dominated by large publishers, make your review count.

Thank you,
Robert Nathan

Made in the USA
San Bernardino, CA
08 September 2019